Lecture Notes in Physics

Springer
Berlin
Heidelberg
New York
Barcelona
Hong Kong
London
Milan
Paris
Tokyo

Physics and Astronomy

ONLINE LIBRARY

http://www.springer.de/phys/

Editorial Policy

The series *Lecture Notes in Physics* (LNP), founded in 1969, reports new developments in physics research and teaching -- quickly, informally but with a high quality. Manuscripts to be considered for publication are topical volumes consisting of a limited number of contributions, carefully edited and closely related to each other. Each contribution should contain at least partly original and previously unpublished material, be written in a clear, pedagogical style and aimed at a broader readership, especially graduate students and nonspecialist researchers wishing to familiarize themselves with the topic concerned. For this reason, traditional proceedings cannot be considered for this series though volumes to appear in this series are often based on material presented at conferences, workshops and schools (in exceptional cases the original papers and/or those not included in the printed book may be added on an accompanying CD ROM, together with the abstracts of posters and other material suitable for publication, e.g. large tables, colour pictures, program codes, etc.).

Acceptance

A project can only be accepted tentatively for publication, by both the editorial board and the publisher, following thorough examination of the material submitted. The book proposal sent to the publisher should consist at least of a preliminary table of contents outlining the structure of the book together with abstracts of all contributions to be included.

Final acceptance is issued by the series editor in charge, in consultation with the publisher, only after receiving the complete manuscript. Final acceptance, possibly requiring minor corrections, usually follows the tentative acceptance unless the final manuscript differs significantly from expectations (project outline). In particular, the series editors are entitled to reject individual contributions if they do not meet the high quality standards of this series. The final manuscript must be camera-ready, and should include both an informative introduction and a sufficiently detailed subject index.

Contractual Aspects

Publication in LNP is free of charge. There is no formal contract, no royalties are paid, and no bulk orders are required, although special discounts are offered in this case. The volume editors receive jointly 30 free copies for their personal use and are entitled, as are the contributing authors, to purchase Springer books at a reduced rate. The publisher secures the copyright for each volume. As a rule, no reprints of individual contributions can be supplied.

Manuscript Submission

The manuscript in its final and approved version must be submitted in camera-ready form. The corresponding electronic source files are also required for the production process, in particular the online version. Technical assistance in compiling the final manuscript can be provided by the publisher's production editor(s), especially with regard to the publisher's own Latex macro package which has been specially designed for this series.

Online Version/ LNP Homepage

LNP homepage (list of available titles, aims and scope, editorial contacts etc.):
http://www.springer.de/phys/books/lnpp/
LNP online (abstracts, full-texts, subscriptions etc.):
http://link.springer.de/series/lnpp/

F. Courbin D. Minniti (Eds.)

Gravitational Lensing: An Astrophysical Tool

Springer

Editors

Frédéric Courbin
Institut d'Astrophysique
et de Géophysique
Université de Liège
Allée du 6 août 17, Bat B5C,
Liège 1
Belgium

Dante Minniti
Pontificia Universidad Católica
Departamento de Astronomia
Av. Vicuña Mackenna 4860
Casilla 306, Santiago 22
Chile

Cover Picture: (see contribution by D. Wittmann in this volume)

Cataloging-in-Publication Data applied for

A catalog record for this book is available from the Library of Congress.

Bibliographic information published by Die Deutsche Bibliothek

Die Deutsche Bibliothek lists this publication in the Deutsche Nationalbibliografie;
detailed bibliographic data is available in the Internet at http://dnb.ddb.de

ISSN 0075-8450
ISBN 978-3-540-44355-1 Springer-Verlag Berlin Heidelberg New York

Springer-Verlag Berlin Heidelberg New York
a member of BertelsmannSpringer Science+Business Media GmbH

http://www.springer.de

© Springer-Verlag Berlin Heidelberg 2002

Typesetting: Camera-ready by the authors/editor
Camera-data conversion by Steingraeber Satztechnik GmbH Heidelberg
Cover design: *design & production*, Heidelberg

Printed on acid-free paper
SPIN: 10894702   54/3141/du - 5 4 3 2 1 0

Preface

All bodies are influenced by gravity in the same way, independent of their mass. In fact, even bodies with no mass are affected by gravity, which acts as any other acceleration vector, following Einstein's "equivalence principle" between gravity and inertial forces. This simple consequence of Einstein's principle yielded the first observational confirmation of the theory of general relativity, with the observation of the apparent displacement of stars seen near the solar limb during a total solar eclipse. This early observation of the phenomenon of *gravitational lensing* marked the beginning of what has now evolved into its own field of astrophysics. Gravitational lensing has even evolved into several sub-fields of astrophysics, and consists of a mature topic studied in detail as a natural phenomenon in itself. It is used to tackle astrophysical problems from a new angle.

Gravitational lensing is starting to be sufficiently well understood that it can be *applied* to other astrophysical areas and can help us to address scientific questions that would otherwise be left without any answer. We have tried to reflect this in the present book, as was done in the selection of topics at the "Dark Matter and Gravitational Lensing" workshop (held in July 2000 in San Pedro de Atacama, Chile) where the writing of the book was initiated. Each chapter covers a "sub-field" of gravitational lensing, with the aim of: describing in a very simple way the basics of the theory, reviewing the most recent developments, and reviewing some of the applications foreseen in the near future.

An introduction to the basics of lens modeling is given in the context of quasar lensing, which is the oldest sub-field of gravitational lensing. The emphasis is put on the cosmological applications, such as the determination of the Hubble parameter H_0. Thanks to the progress with instrumentation and the development of large telescopes working at high angular resolution, the weakest effects of gravitational lensing can now be detected. The so-called "weak gravitational lensing" is the topic of the second chapter. It describes how to weigh galaxy clusters and how to map the – invisible – large scale structures of the Universe thanks to the distortion they produce on very distant objects. Weak lensing has been recently extended to the statistical study of the shape of the dark halo in individual galaxies: "galaxy-galaxy lensing" is the subject of the third chapter. Finally, gravitational lensing is starting to be intensely studied at millimeter wavelengths, and is often used as a natural telescope to unveil faint sources otherwise inaccessible. The last chapter gives a broad overview of the applications of gravitational lensing, at these wavelengths, that are just starting to be explored.

Liège, Belgium   *Frédéric Courbin*
Santiago de Chile, Chile   *Dante Minniti*
August, 2002

Table of Contents

List of Contributors

Danielle Alloin
European Southern Observatory
Casilla 19001, Santiago 19, Chile
dalloin@eso.org

Roger D. Blandford
CalTech
1200 East California Boulevard
Pasadena, CA 91125, USA
rdb@tapir.caltech.edu

Tereasa G. Brainerd
Boston University,
Department of Astronomy,
Boston, MA 02215, USA
tgb@firedrake.bu.edu

Frédéric Courbin
Universidad Católica de Chile
Av. Vicuña Mackenna 4860
Casilla 306, Santiago 22, Chile
fcourbin@astro.puc.cl

Prasenjit Saha
Astronomy Unit
School of Mathematical Sciences
Queen Mary and Westfield College
London E1 4NS, UK
P.Saha@qmw.ac.uk

Paul L. Schechter
Center for Space Research
Massachusetts Institute of Technology
70 Vassar Street, Cambridge
MA 02139, USA
schech@achernar.mit.edu

Tommy Wiklind
Onsala Space Observatory
Onsala 43992, Sweden
tommy@oso.chalmers.se

David Wittman
Bell Laboratories
Lucent Technologies, Room 1E-414
700 Mountain Avenue
Murray Hill, NJ 07974, USA
wittman@physics.bell-labs.com

1 Quasar Lensing

Frederic Courbin[1,2], Prasenjit Saha[3], and Paul L. Schechter[4]

[1] Pontificia Universidad Católica de Chile, Av. Vicuña Mackenna 4860,
   Departamento de Astronomia y Astrofisica, Casilla 306, Santiago 22, Chile
[2] Université de Liège, Institut d'Astrophysique et de Géophysique,
   Allée du 6 août, Bat. B5C, Liège 1, Belgium
[3] Astronomy Unit, School of Mathematical Sciences
   Queen Mary and Westfield College, London E1 4NS, UK
[4] Massachusetts Institute of Technology,
   70 Vassar Street, Cambridge, MA 02139, USA

Abstract. Massive structures, such as galaxies, act as strong gravitational lenses on background sources. When the background source is a quasar, several lensed images are seen, as magnified or de-magnified versions of the same object. The detailed study of the image configuration and the measurement of "time-delays" between the images yield estimates of the Hubble parameter H_0. We describe in a simple way the phenomenon of strong lensing and review recent progress made in the field, including microlensing by stars in the main lensing galaxy.

1.1 Concepts

1.1.1 The Formation of Multiple Images

There are several ways of understanding the effect of gravity on light in the context of lensing. We start with an approach which lends itself particularly well to pictorial representation.

Wavefronts. A schematic wavefront is illustrated in Fig. 1.1. Spreading outwards from a point source, the wavefront is initially spherical. But as it passes through the gravitational field of the lens the wavefront gets delayed and bent; we can interpret this effect as a slowing-down of light by a gravitational field, usually called the Shapiro time delay [124]. Where the wavefront crosses an observer, they see an image in the direction normal to the wavefront, and images will be (de)magnified and/or distorted according to how curved the wavefront is as it crosses the observer. If the lens is strong enough, the wavefront can fold in on itself, producing multiple images. If moreover the source is variable, different images will show that variability with time delays proportional to the spacing between these folds, i.e., the cosmological distance scale.

It is possible to make the above explanation quantitative within the wavefront picture [104,55], but for calculations that is usually not the most convenient route. Notice that the wavefront picture has a single source and multiple observers, whereas astrophysical problems generally involve multiple sources and a single observer. So calculations are easier if we use a relative of the wavefront called the arrival-time surface [16,91].

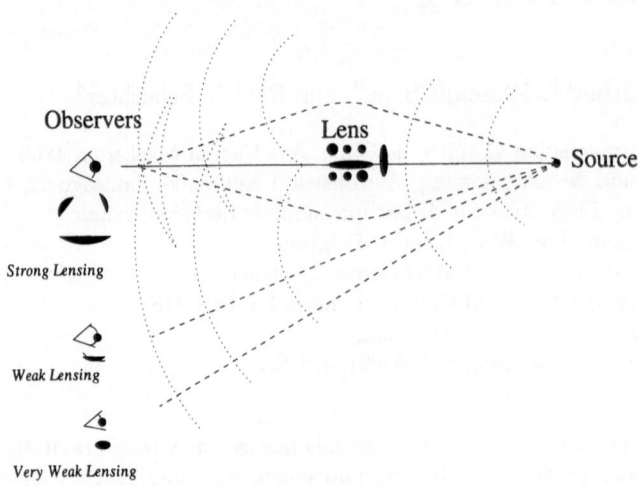

Fig. 1.1. Schematic illustration [1] of the wavefront and the different regimes of lensing. Lensed quasars fall in the strong lensing regime; the other regimes are important in lensing by clusters of galaxies.

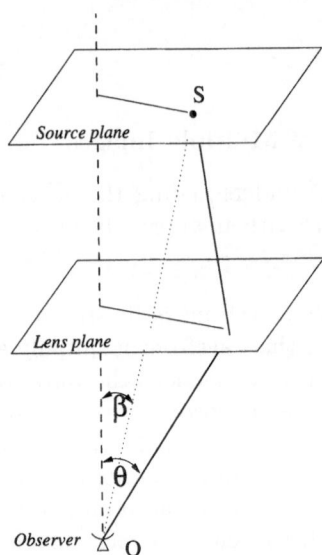

Fig. 1.2. Illustration of a virtual light ray: $\vec{\beta}$ is the unlensed sky position of the source, and $\vec{\theta}$ is its apparent position. In the text, we use D_L, D_S, and D_{LS} for angular diameter distances from observer to lens, observer to source, and lens to source.

Arrival Times. Consider Fig. 1.2: in the usual astrophysical approximation of small angles and thin lenses, this figure shows a virtual light ray getting deflected by the lens and reaching the observer from direction $\vec{\theta}$, the source being at angular position $\vec{\beta}$. (Vector signs denote 2D angles on the sky.) The arrival time is the light travel time—with irrelevant constants discarded—of such a virtual ray as a function of $\vec{\theta}$, with $\vec{\beta}$ held fixed. It has two contributions: a 'geometrical' part and a 'gravitational' part [16]. The geometrical part is simply the difference between the continuous and dotted paths in Fig. 1.2, and is given by

$$t_{geom}(\vec{\theta}) = \tfrac{1}{2}(1+z_L)\frac{D_L D_S}{cD_{LS}}(\vec{\theta}-\vec{\beta})^2, \tag{1.1}$$

where z_L is the lens redshift and the D factors are angular diameter distances as shown in Fig. 1.2. The gravitational part is the Shapiro time delay in a gravitational field from general relativity, and depends on the surface density $\Sigma(\vec{\theta})$ of the lens. A concise way of writing the Shapiro delay is

$$t_{Shapiro}(\vec{\theta}) = (1+z_L)\frac{8\pi G}{c^3}\nabla^{-2}\Sigma(\vec{\theta}). \tag{1.2}$$

Here ∇^{-2} denotes the inverse of a 2D Laplacian with respect to $\vec{\theta}$,[1] and $\nabla^{-2}\Sigma(\vec{\theta})$ is some sort of 2D potential.

Putting (1.1) and (1.2) together we have the arrival time $t(\vec{\theta})$ in full:

$$t(\vec{\theta}) = \tfrac{1}{2}(1+z_L)\frac{D_L D_S}{cD_{LS}}(\vec{\theta}-\vec{\beta})^2 - (1+z_L)\frac{8\pi G}{c^3}\nabla^{-2}\Sigma(\vec{\theta}). \tag{1.3}$$

From Fermat's principle, real light rays take paths that make the arrival time stationary. Thus the condition for images is

$$\vec{\nabla}t(\vec{\theta}) = 0. \tag{1.4}$$

Equation (1.3) looks formidable, but it will become much less so once we introduce some scales.

Some Scales. Consider a point-mass lens and a point source along the same line of sight, i.e., $\vec{\beta}=0$ and $\Sigma(\vec{\theta})=M\delta(\vec{\theta})$. The arrival time then becomes

$$t(\vec{\theta}) = \tfrac{1}{2}(1+z_L)\frac{D_L D_S}{cD_{LS}}\theta^2 - (1+z_L)\frac{4G}{c^3}\ln\theta, \tag{1.5}$$

since $\nabla^{-2}\delta(\vec{\theta}) = \ln\theta/(2\pi)$, and there is a minimum at $\theta = \theta_E$ where

$$\theta_E^2 = \frac{4GM}{c^2}\frac{D_{LS}}{D_L D_S}. \tag{1.6}$$

---

[1] By this we mean an operator that solves Poisson's equation in 2D. Thus, if $\nabla^2 f(\vec{\theta}) = g(\vec{\theta})$, we write $f(\vec{\theta}) = \nabla^{-2}g(\vec{\theta})$. The explicit form of the inverse Laplacian is as an integral

$$f(\vec{\theta}) = \int \ln|\vec{\theta}-\vec{\theta}'|\,g(\vec{\theta}')\,d^2\vec{\theta}'$$

but we will not need it in this article.

This corresponds to a ring image, called an Einstein ring, and θ_E is called the Einstein radius. If the source is much further than the lens

$$\theta_E \simeq 0.1 \, \text{arcsec} \times \left[\frac{M \, \text{in} \, M_\odot}{D_L \, \text{in} \, \text{pc}} \right]^{\frac{1}{2}}. \tag{1.7}$$

The combination of a point lens and colinear source is very improbable, but the Einstein radius is a very useful concept, for two reasons. First, even if there is no Einstein ring in a multiple-image system, the image separation still tend to be of order θ_E. Secondly, the Einstein radius also supplies a scale for Σ, by the following argument.

From the two-dimensional analog of Gauss's flux law, for any circular mass distribution $\Sigma(\theta)$, $\vec{\nabla} t(\vec{\theta})$ will depend only on the enclosed mass. So not just a point mass, but any circular distribution of the mass M, will produce an Einstein ring from a colinear source, *provided it fits within* θ_E. The condition of a mass fitting into its own Einstein radius is known as 'compactness'. And because from (1.6) the area within an Einstein radius is itself proportional to the mass, compactness is equivalent to the density exceeding some critical density . Working out the algebra we easily get this critical density[2] to be

$$\Sigma_{\text{crit}} = \frac{c^2}{4\pi G} \frac{D_L D_S}{D_{LS}}. \tag{1.8}$$

From (1.3) we can also define a time scale

$$T_0 = (1 + z_L) \frac{D_L D_S}{c D_{LS}}, \tag{1.9}$$

which is of order the light travel time, or a Hubble time in cosmological situations. The interesting time scale in lensing, however, is not T_0, but

$$T_0 \times \langle \text{image separations} \rangle^2, \tag{1.10}$$

being the scale of arrival-time *differences* between images. We will meet the latter presently, in the approximate (1.14).

The Arrival-Time Surface. Using the scales introduced above, we can render dimensionless the arrival time (1.3),

$$\tau(\vec{\theta}) = \tfrac{1}{2}(\vec{\theta} - \vec{\beta})^2 - 2\nabla^{-2}\kappa(\vec{\theta}). \tag{1.11}$$

Here the scaled arrival time τ and the scaled surface density κ (also called convergence) are both dimensionless. The last term in (1.11) is called the lens or projected potential

$$\psi(\vec{\theta}) \equiv 2\nabla^{-2}\kappa(\vec{\theta}). \tag{1.12}$$

---

[2] In this article the units of Σ are $M_\odot$ arcsec^{-2}. Some authors prefer $M_\odot$ kpc^{-2}. This difference of convention means that different authors' equations may differ by factors of D_L or D_L^2.

The physical arrival time and density are

$$t(\vec{\theta}) = \tau(\vec{\theta}) \times T_0, \quad \Sigma(\vec{\theta}) = \kappa(\vec{\theta}) \times \Sigma_{\text{crit}}, \tag{1.13}$$

and the scales are approximately

$$T_0 \simeq h^{-1} z_{\text{L}} (1 + z_{\text{L}}) \times 80 \, \text{days} \, \text{arcsec}^{-2}. \tag{1.14}$$

and

$$\Sigma_{\text{crit}} \simeq h^{-1} z_{\text{L}} \times 1.2 \cdot 10^{11} \, M_{\odot} \, \text{arcsec}^{-2}, \tag{1.15}$$

where h is the Hubble constant in units of 100 km/s/Mpc.

The scaled arrival time $\tau(\vec{\theta})$ in (1.11), visualized as a surface, is called the arrival-time surface. Much of lensing theory is effectively the study of the arrival-time surface and its derivatives, as we see below.

Note that although the wavefront and the arrival time surface look similar and indeed are closely related [91], they are not quite the same thing. The wavefront is a surface in real space whereas the arrival time surface is in $(\vec{\theta}, \tau)$ space and thus a little more abstract.

Images and Magnification. The condition for images, from Fermat's principle and following (1.4) is

$$\vec{\nabla} \tau(\vec{\theta}) = 0, \quad \text{or} \quad \vec{\beta} = \theta - \vec{\nabla} \psi. \tag{1.16}$$

The latter form is called the lens equation. Its interpretation is that the observer sees an image wherever the arrival-time surface has a minimum, maximum, or saddle point. Then consider the second derivative of $\tau(\vec{\theta})$, or curvature of the arrival-time surface. We have

$$\vec{\nabla}\vec{\nabla}\tau(\vec{\theta}) = \mathbf{1} - \vec{\nabla}\vec{\nabla}\psi(\vec{\theta}), \tag{1.17}$$

a 2D tensor. (The bold-face $\mathbf{1}$ denotes an isotropic tensor—identity matrix in component notation.) Meanwhile, taking the gradient of the lens equation (1.16) gives

$$\vec{\nabla}\vec{\beta} = \mathbf{1} - \vec{\nabla}\vec{\nabla}\psi(\vec{\theta}). \tag{1.18}$$

The curious term $\vec{\nabla}\vec{\beta}$ expresses how much source-plane displacement is needed to produce a given small image displacement; i.e., the inverse of magnification [3]. Equation (1.18) tells us that magnification is a 2D tensor, and depends on $\vec{\theta}$ but not $\vec{\beta}$; let us write magnification as $\mathbf{M}$. Comparing the last two equations we have

$$\mathbf{M}^{-1} = \vec{\nabla}\vec{\nabla}\tau(\vec{\theta}). \tag{1.19}$$

Equation (1.19) means that the curvature of the arrival-time surface is the inverse of the magnification. Thus, broad low hills and shallow valleys in the arrival-time

---

[3] An alternative notation, $\partial\vec{\beta}/\partial\vec{\theta}$, reminds one of this physical interpretation.

surface correspond to highly magnified images; needle-sharp peaks or troughs correspond to images demagnified into unobservability.

By curvature, we mean a tensor curvature, which depends on directions: $\mathbf{M}$ and $\mathbf{M}^{-1}$ are symmetric 2D tensors, so their components form 2×2 matrices. In particular, we have

$$\mathbf{M}^{-1} = \begin{pmatrix} 1 - \partial^2\psi/\partial\theta_x^2 & -\partial^2\psi/\partial\theta_x\partial\theta_y \\ -\partial^2\psi/\partial\theta_x\partial\theta_y & 1 - \partial^2\psi/\partial\theta_y^2 \end{pmatrix}. \tag{1.20}$$

Comparing (1.11) and (1.19 or 1.20) we see that the trace of $\mathbf{M}^{-1}$ must be $2(1 - \kappa)$. Thus κ, originally defined as the surface density in suitable units, also has the interpretation of an isotropic magnification. Accordingly, κ is known as the convergence. The traceless part of $\mathbf{M}^{-1}$ is called the shear and its magnitude is denoted by γ; it changes the shape of an image but not its size. In full, we have

$$\mathbf{M}^{-1} = (1 - \kappa) \begin{pmatrix} 1 & 0 \\ 0 & 1 \end{pmatrix} - \gamma \begin{pmatrix} \cos 2\phi & \sin 2\phi \\ \sin 2\phi & -\cos 2\phi \end{pmatrix} \tag{1.21}$$

where ϕ denotes the direction of the shear. Note that any symmetric 2×2 matrix can be written in the form (1.21). All we have done here is interpret κ and γ.

The determinant

$$|\mathbf{M}| = [(1 - \kappa)^2 - \gamma^2]^{-1} \tag{1.22}$$

defines a scalar magnification, or ratio of image area to source area for an infinitesimal source.

Surface brightness is conserved by lensing. Although we will not prove it here, this is a consequence of the fact that the lens equation is a gradient map. Magnification changes only angular sizes and shapes on the sky. Thus a constant surface brightness sheet stays a constant brightness sheet when lensed. (Were this not the case, the microwave background would get wildly lensed by large scale structure.) However, an unresolved source will have its brightness amplified according to (1.22).

Saddle-Point Contours, Critical Curves, Caustics. The equations (1.11) for the arrival-time surface, (1.16) for the image positions, and (1.19) for the magnification are elegant, but they do not give us much intuition for the shape of the arrival-time surface, the possible locations of images, and the likely magnification in real systems that we might observe. To gain some intuition, it is very useful to introduce [16] three special curves in the image and source planes.

Consider the arrival-time surface and contours of constant τ. In the absence of lensing $\tau(\vec{\theta})$ is a parabola, and the image is at its minimum, or $\vec{\theta} = \vec{\beta}$. For a small lensing mass, the shape changes slightly from being a parabola and the minimum moves a little. But for large enough mass, a qualitative change occurs, in that a contour becomes self-crossing. There are two ways in which a self-crossing can develop: as a kink on the outside of a contour line, or a kink on the inside. These are illustrated in Fig. 1.3. The outer-kink type is a lemniscate and the inner-kink type a limaçon. With the original contour having enclosed a minimum, a lemniscate produces another minimum, plus a saddle-point at the self-crossing, while a limaçon produces a new maximum plus a saddle point. (The previous sentence remains valid if we interchange the words 'maximum' and 'minimum'.) The process of contour self-crossing can then repeat around any of the new maxima and minima, producing more and more new images, but always satisfying

$$\text{maxima} + \text{minima} = \text{saddle points} + 1. \tag{1.23}$$

The self-crossing or saddle-point contours form a sort of skeleton for the multiple-image system. Lensed quasars characteristically have one of two configurations: double quasars have a limaçon, while quads have a lemniscate inside a limaçon, as in the rightmost part of Fig. 1.3. Both cases have one maximum (marked 'H' in the figure), which will be located at the center of the lensing galaxy. Since galaxies tend to have sharply-peaked central densities, the arrival-time surface at the maximum will be sharply peaked as well; the corresponding image is highly demagnified and (almost) always unobservable. Thus lensed quasars are doubles or quads: an incipient third or fifth image hides at the center of the lensing galaxy.

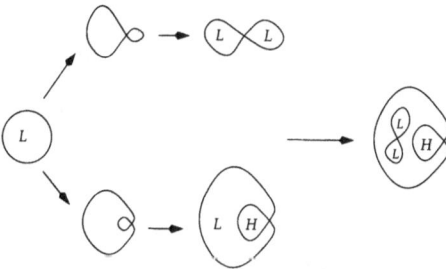

Fig. 1.3. Multiple images via saddle-point contours in the arrival-time surface. Here L marks minima and H marks maxima.

Critical curves are curves on the image plane where the magnification is infinite. More formally, they are curves where $\mathbf{M}^{-1}$ has a zero eigenvalue. From the definition (1.19) it is clear that at minima of τ both eigenvalues of $\mathbf{M}^{-1}$ will be positive, at maxima both eigenvalues will be negative, and at saddle points one eigenvalue will be positive and one negative. Thus critical curves separate regions of the image plane that allow minima, saddle points, and maxima.

If we map critical curves to the source place via the lens equation (1.16) we get caustic curves. Caustics separate regions on the source plane that give rise to different numbers of images.

We discuss examples of saddle-point contours, critical curves, and caustics in the next section.

1.1.2   An Illustrative Macro-model

We have already met the point lens, which in dimensionless form has lens potential

$$\psi(\vec{\theta}) = \theta_E \ln \theta \tag{1.24}$$

where θ_E is effectively a parameter expressing the total mass. Solving the lens equation, we see that images are at

$$\vec{\theta} = \tfrac{1}{2}\left(\beta \pm \sqrt{\beta^2 + 4\theta_E^2}\right)\hat{\beta}, \tag{1.25}$$

where $\hat{\beta}$ denotes a unit vector in the direction of $\vec{\beta}$. The scalar magnification is given by

$$|\mathbf{M}|^{-1} = 1 - \frac{\theta_E^4}{\theta^4}. \tag{1.26}$$

Another commonly used lens is the isothermal lens (so called because of its relation to isothermal spheres in stellar dynamics, and a good zeroth order model for disk-galaxy halos and giant ellipticals — more on this subject in the modeling section); it has $\kappa(\vec{\theta}) = \tfrac{1}{2}\theta_E/\theta$ and lens potential

$$\psi(\vec{\theta}) = \theta_E \theta, \tag{1.27}$$

For $\beta < \theta$ there are two images at

$$\vec{\theta} = \vec{\beta} + \theta_E \hat{\beta}, \qquad \vec{\theta} = \vec{\beta} - \theta_E \hat{\beta} \tag{1.28}$$

and for $\beta > \theta$ the second of these disappears. The constant image-separation in (1.28) is a peculiar feature of the isothermal. The scalar magnification is given by

$$|\mathbf{M}|^{-1} = 1 - \frac{\theta_E}{\theta}. \tag{1.29}$$

Lacking any ellipticity, these lenses by themselves cannot produce quads. But with some added ellipticity, quads and indeed all the main qualitative features of quasar lenses can be reproduced, as we now show.

As an example, consider the potential

$$\psi(\vec{\theta}) = (\theta^2 + \epsilon^2)^{\frac{1}{2}} + \tfrac{1}{2}\gamma\theta^2 \cos(2\phi) \tag{1.30}$$

where ϕ is the polar angle and ϵ and γ are adjustable parameters; ϵ gives the isothermal a non-singular core, and $\gamma > 0$ contributes 'external shear' which in this case amounts to extra lensing mass outside the lens in the y direction. We take $\epsilon = 0.1$ and $\gamma = 0.2$, and then examine what happens for different source positions, through caustics, critical curves, and saddle-point contours. A similar potential, but with the scale and shear orientation adjustable, will be used later (cf. (1.44)) to fit data on observed systems.

Figure 1.4 shows the situation with the source close to the center. The left panel shows what is happening in the source plane, while the middle and right panel show what is happening in the image plane. Several interesting things may be seen.

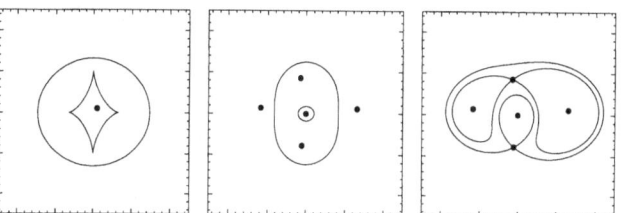

Fig. 1.4. A central quad: one with source near the center. **Left panel:** source positions and caustics; **middle panel:** image positions and critical curves; **right panel:** image positions and saddle-point contours. In this figure, and in Figs. 1.5 to 1.7, the left hand panels (showing the source plane) have a scale half that of the other panels (showing the image plane).

- The two caustic curves in the source plane (left panel) demarcate regions from where a source produces 1, 3, and 5 images. In this case the source is well within the inner caustic, and that results in five images. The other panels shows these five images, along with the critical curves (middle panel) or the saddle-point contours (right panel). But the image near the center is highly demagnified, and observationally such a system would be a quad. Let us call it a 'central quad', to distinguish it from other quads we will see below.
- The two critical lines are maps of the caustics to the image plane, but the *inner* caustic maps to the *outer* critical line. (Also, the long axes of both of these are aligned with the potential.) Recall that critical lines are where an eigenvalue of **M** changes sign. The consequence for this lens is that any image outside both critical curves is a minimum, any image between the critical curves is a saddle point, and any image inside both critical curves is a maximum, all irrespective of the source position. For the current source position we can verify these statements by comparing the middle and right panels.
- The time-ordering of the quad's images is evident from the saddle-point contours—compare with Fig. 1.8.
- The arrival-time contours and the arrangement of the images appear to be squeezed in the y direction. Such squeezing is characteristically along the long axis of the potential, and the images appear pop out along the short axis of the potential.

Figure 1.5 shows the situation with the source is displaced along the long axis of the potential. As the source nears the inner caustic curve, two of the images approach the outer critical curve. We call this configuration a long-axis quad. Two minima and a saddle point are fairly close together, displaced in the same direction as the source, while another saddle point is on the opposite side of the lens center. This 3+1 image arrangement reveals the direction of the source displacement. Meanwhile, as with the core quad, the arrangement of the images is squeezed along the direction of the long axis of the potential. As the source

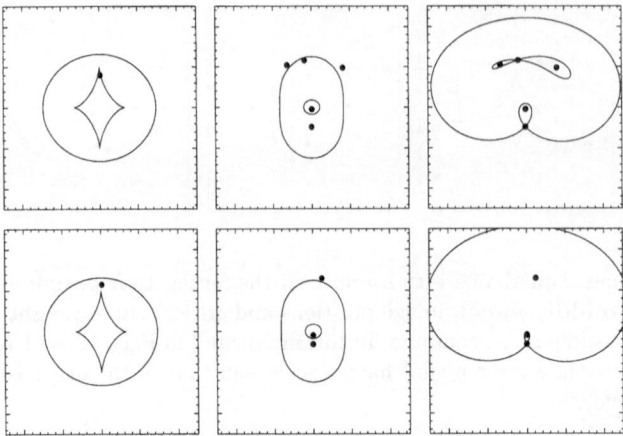

Fig. 1.5. A long axis quad and double. Note how, as the source crosses the diamond caustic, two images merge on the tangential critical line and then disappear.

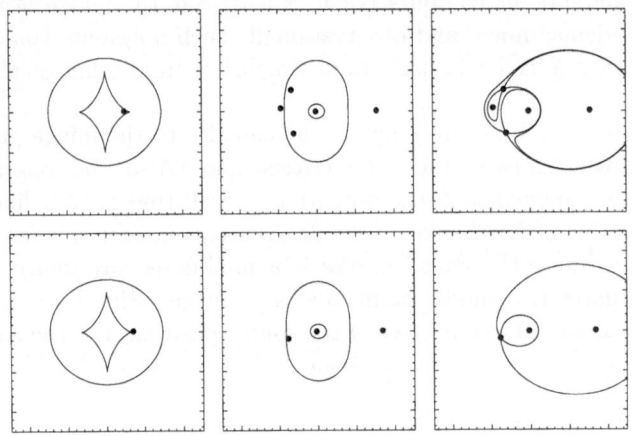

Fig. 1.6. A short axis quad and double.

crosses the inner caustic curve, a minimum and a saddle point merge on the outer critical curve, and then disappear. The system is now a double, which we may call a long-axis double.

Figure 1.6 has the source displaced along the short axis of the potential, producing configurations we call a short-axis quad and a short-axis double. The morphology of a short-axis quad resembles that of the long axis quad, but one can tell them apart. First, one of the four images is far from the others, but this time it is a minimum, not a saddle point. Secondly, the 1+3 image arrangement indicates direction of the source displacement, and it is perpendicular to the long axis of the potential which can be inferred from the squeezing of the image arrangement. Moving the source outside the inner caustic again causes two images to merge, leaving a short-axis double. The morphology of a short axis-double is the same as that of a long-axis double.

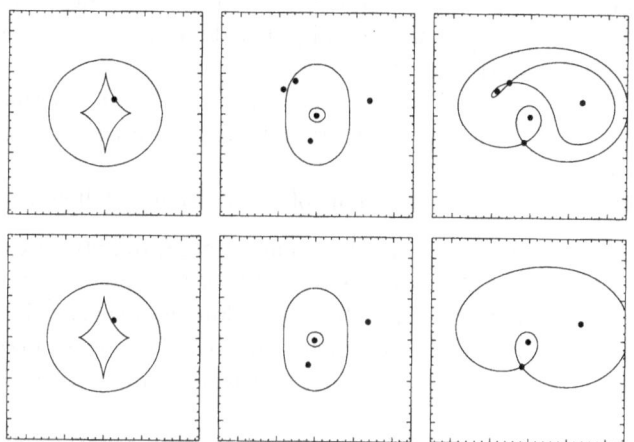

Fig. 1.7. An inclined quad and double.

Figure 1.7 has the source displaced obliquely to the potential, producing what we call an inclined quad and an inclined double. These are more common than the long and short-axis types, and easily distinguished because of their asymmetry.

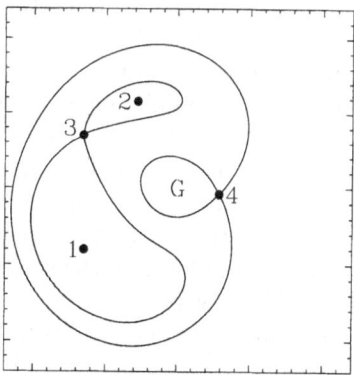

Fig. 1.8. Saddle point contours in a generic quad. Images 1 and 2 are minima, 3 and 4 are saddle points; the fifth image would be a maximum at the galaxy's centre G.

Examining the saddle-point contours in Figs. 1.4 to 1.7, the order of arrival times of the images is nearly always evident. We can summarize image-ordering in quads in the following simple rules, illustrated in Fig. 1.8: (i) Images 1 and 2 are opposite in Position Angle (PA), (ii) 3 and 4 are opposite in PA, (iii) 1 is the furthest or nearly the furthest from the lens centre, (iv) 4 is the furthest or nearly the furthest from the lens centre, (v) if there are a nearly merging pair, they are 2 and 3. For some cases it is not possible to decide between 1 and 2, but otherwise there is never an ambiguity. For doubles, time ordering is trivial: the image further from the galaxy arrives first.

With a little practice, it is easy to sketch the saddle-point contours (including image ordering), and from there the critical curves and caustics, of any quasar lens just from the morphology.

We may summarize the conclusions of this section as follows:

- From the morphology of a quad, it may be immediately recognized as one of (i) central, (ii) long axis, (iii) short axis, or (iv) inclined; doubles may be recognized as (i) axis, or (ii) inclined, but long and short axis doubles need more information to distinguish. The 'axis' in each case is of course the axis of the potential, including any external shear; so morphology already gives some idea mass distribution.
- Morphology of quads or doubles also reveals the time-ordering of images.

1.1.3 Lenses Within Lenses: Microlensing

Stars comprise an appreciable fraction of the mass in lensing galaxies. These stars produce small scale fluctuations in lens potentials which will be seen to have substantial effects on the magnifications of quasar images.

Random Star Fields. Suppose that light from a source passes through a screen of N equal point masses with random positions, and that the Einstein rings of individual masses are very small compared to the mean spacing between them. The delay equation (1.3) consists of a single geometric term and a great many Shapiro terms. Each Shapiro term produces three stationary points: a singular and completely demagnified maximum at the angular position of the point mass, a significantly demagnified saddle point close to the point mass and on the side opposite the source position, and a minimum not far from the unperturbed source position. The minimum will be the only non-negligible image.

Were the same stars to increase in mass without changing position, the saddle points would move further away from the stars, increasing in brightness. As their masses continue to increase, close pairs will create new saddles between them. For each new ridge, a new valley will form on the side furthest from the source. Just after the formation of this new pair of images, the curvature along the line connecting them is very small and they are very highly magnified. As the masses continue to increase, the images separate and grow fainter, though the new minimum will never be fainter than the unmagnified source.

Thus the number of images increases from $N + 1$ to $N + 3$ to $N + 5$ and so forth. If we lacked the resolution to see the individual images, but only the combined light, we would find that for the most part the combined brightness increases steadily, but with bright flashes as new pairs of images are created. At any time our star field would have some average surface density and an associated dimensionless convergence, κ. For an ensemble of such sources placed randomly behind such a screen, we would expect an average scalar magnification of $(1 - \kappa)^{-2}$ (see (1.22)), but there would be fluctuations depending upon the accidents of source position. Additional images begin to appear (in the absence of external shear) when κ approaches unity.

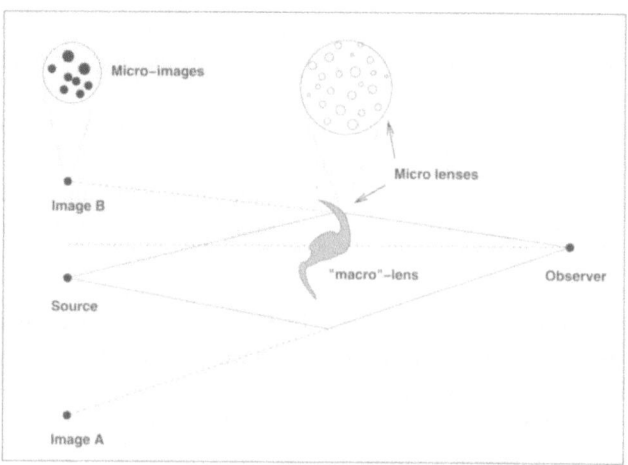

Fig. 1.9. Schematic representation of microlensing by stars in a doubly imaged system. In this example, the unresolved "sea" of stars in the main lensing galaxy is responsible for "microlensing" of one of the quasar images.

The general phenomenon of the amplification of unresolved images by stars (or other point masses) in intervening galaxies is termed microlensing. The situation is illustrated schematically in Fig. 1.9. The large numbers of highly demagnified saddle points are not shown.

Mandatory Microlensing. In the thought experiment of the preceding subsection, additional positive parity images (minima) and their accompanying saddle points formed when κ approached unity. The average density interior to the Einstein ring of an isolated microlens is just the critical density, with $\kappa \equiv 1$. The criterion for substantial microlensing is therefore $\kappa \sim 1$.

Now let us suppose that the galaxy lensing a multiply imaged quasar is comprised entirely of point masses. The average surface density interior to the galaxy's Einstein ring is exactly the critical density. Unless the galaxy is very highly concentrated, the surface density at the Einstein ring must be a substantial fraction of the critical density – one half in the case of an isothermal lens. The covering factor of the microlenses' Einstein rings must therefore be a substantial fraction of unity. Thus microlensing must be important, *if* the galaxy is comprised entirely of point-like objects.

Microlensing will be important only if the Einstein rings of the particles comprising the galaxy are larger than the projection of the source onto the sky. There are two ways in which this might fail to occur for lensed quasars. First, the source might be large compared to the the Einstein rings of the galaxy's stars. Second, most of the mass in the galaxy might be in particles with masses very much smaller than that of a star, as we suspect would be the case for dark matter. Our understanding of quasar sources and the distribution of dark matter within galaxies is as yet so limited than we cannot say with certainty whether microlensing should or should not be expected. As we shall see later there is considerable observational evidence that the conditions for microlensing are met in at least some lensed quasars.

It should be noted, by contrast, microlensing of sources in the Magellanic Clouds by stars (and dark objects) in the halo of our Milky Way is an exceedingly rare event. The covering factor for halo object Einstein rings is at most 10^{-6}. The largest source of this difference is the small distance to the Clouds and the correspondingly large value of Σ_{crit}.

Static and Kinetic Microlensing. In the above *Gedanken* experiment neither the source nor stars were moving. Imagine a symmetric lens which forms two quasar images exactly opposite each other. The images pass through regions of identical surface density and shear, and would, in the absence of microlensing, undergo the same magnification. But since they pass through different random star fields, they suffer different amounts of microlensing. The magnifications predicted from the global galaxy potential would be only approximate – one would have to take into account the local fluctuations. Static microlensing produces "errors" in the predicted fluxes.

Imagine further that the quasar consists of two components, one smaller than the typical size of stellar Einstein rings and the other larger. The smaller component would be microlensed but the larger component would not.

The motions of the source and the microlensing pattern add an additional complication. Taking the microlens positions to be fixed, as the source moves the microlensing will change. To order of magnitude, the source must move an amount equal to the Einstein radius of the microlens to produce a substantial change. If the stars are moving, they must move an amount comparable to the sizes of their Einstein radii to produce substantial changes. The temporal changes in the brightness of an unresolved source are the result of such kinetic microlensing.

Microlensing Caustics. As described in Sect. 1, critical curves are the locii in the image plane along which pairs of images merge or are created as one varies the position of a background source. The scalar magnification is infinite along the critical curves. This property suggests a relatively straightforward computational scheme for identifying caustics, which are the locii in the source plane which produce images on the critical curves. Given a set of (random) microlens

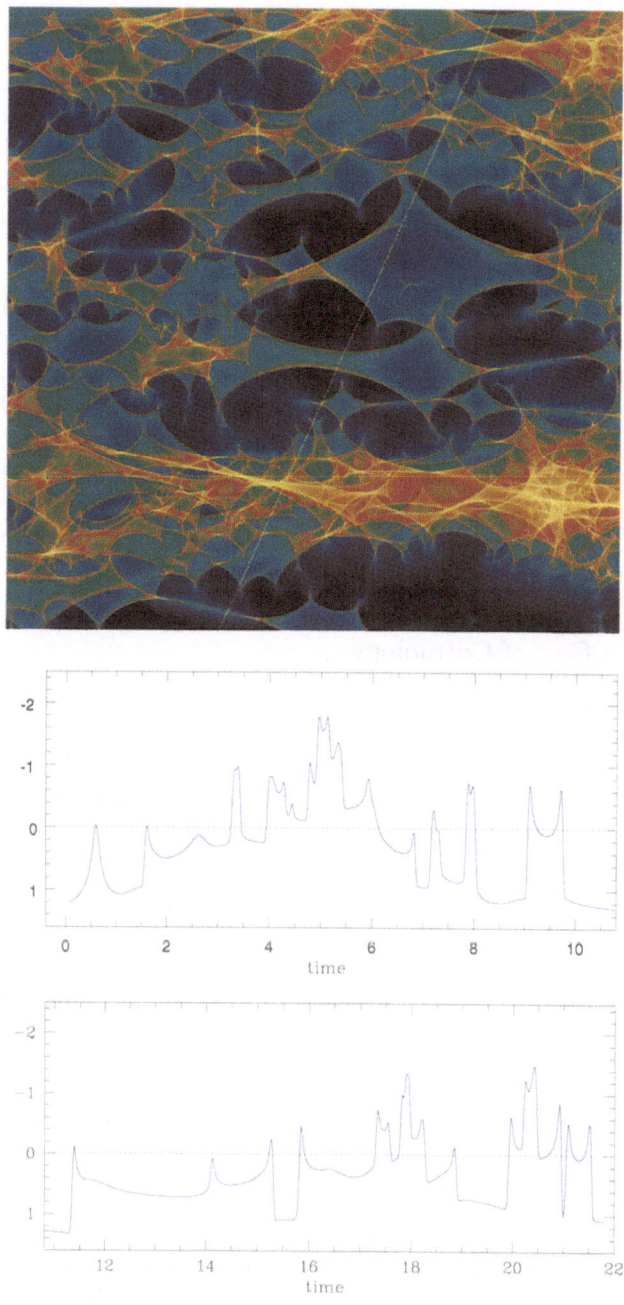

Fig. 1.10. Top: network of micro-caustics in a lensing galaxy. The local convergence κ is 0.5 and the shear γ is 0.6. The bright regions correspond to high magnification while the darker ones show de-magnification. **Bottom:** predicted light curve when a source crosses the caustics along the straight line in the top panel. The time scale is arbitrary (Figure courtesy Joachim Wambsganss) (see Color Plate).

positions, one projects rays back from the observer uniformly in solid angle. These land with high (low) surface density in regions of high (low) magnification and the caustics readily emerge when one plots a "spot diagram" for these rays. Such a plot also allows rapid computation of kinetic microlensing light curves for a moving source – one simply takes linear cuts through the source plane spot diagram. The magnification is proportional to the local density of spots. Figure 1.10 shows such a plot, with a predicted light curve when a source crosses a network of caustics.

Quantitative Microlensing. Microlensing is fundamentally statistical in nature. It has been surprisingly resistant to analytic techniques, and most quantitative work has been carried out via simulations. These have shown [135,73] that fluctuations of a magnitude or more are possible for highly magnified images. Moreover saddlepoints behave differently from minima, with larger fluctuations for the former than for the latter [147,118]. Among the few interesting analytic results are an exact expression for the magnification probability distribution at high magnification [122], and an expression for the mean number of positive parity microimages (minima) as a function of κ.

1.1.4   The Effect of Cosmology

The main observables in lensing, image positions and magnifications, are all dimensionless; only time delays are dimensional. The effect of cosmology is to set the scale of time delays, and we can think of it as setting T_0, the time scale in (1.9). Cosmology really enters only through the angular-diameter distances, so fixing T_0 also fixes the other important scale, Σ_{crit}.

The time scale has a dependence of the form

$$T_0 = h^{-1} z_{\mathrm{L}} (1 + z_{\mathrm{L}}) \times \langle \text{weak function of } z_{\mathrm{L}}, z_{\mathrm{S}}, \Omega_0, \Omega_\Lambda \rangle \qquad (1.31)$$

and is analytic [37] but messy, so we do not reproduce it here. Instead we illustrate it in Fig. 1.11 for some cosmologies. It is worth remarking that

- $T_0 \propto h^{-1}$ exactly;
- for $z_{\mathrm{S}} \gg z_{\mathrm{L}}$ the approximation (1.14) applies.
- for the same h, an Einstein-de-Sitter cosmology gives large T_0, an open cosmology gives small T_0, with the currently favored flat Λ-cosmology being intermediate; but the differences are small.

The simple dependence on h make it attractive to use time delays to try and measure h. One can even imagine putting in several time delays on a sort of Hubble diagram to try and constrain Ω_0, Ω_Λ. Both these ideas are due to Refsdal [105,106].

1.1.5   Degeneracies

Lensed images correspond to minima, saddle points, and maxima of the arrival-time surface; the rest of the arrival-time surface is unobservable. Thus, lensing

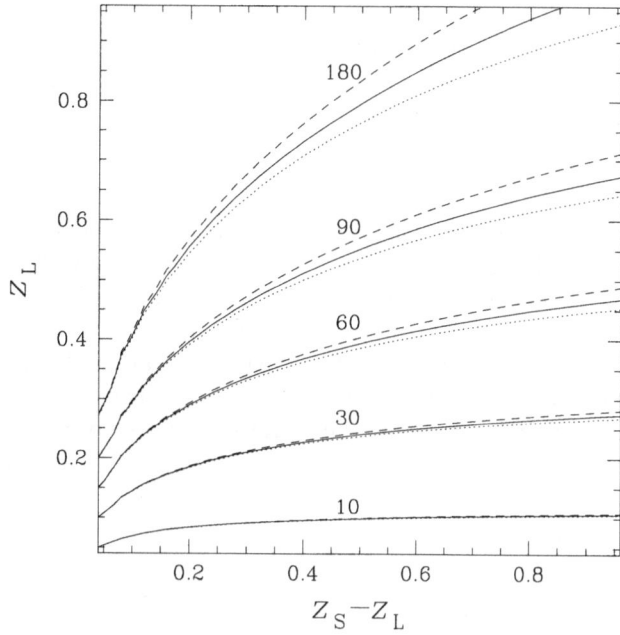

Fig. 1.11. Contour plots of T_0 as a function of $z_S - z_L$ and z_L. The labels are in units of h^{-1}days arcsec^{-2}. The dashed curves are for $\Omega_0 = 1, \Omega_\Lambda = 0$, the solid curves are for $\Omega_0 = 0.3, \Omega_\Lambda = 0.7$, and the dotted curves are for $\Omega_0 = 0.1, \Omega_\Lambda = 0$.

observables do not uniquely specify a lens; another lens that preserves $\tau(\vec{\theta})$ and its derivatives at image positions but changes them elsewhere will produce exactly the same lensing data. In this sense, lenses are subject to degeneracies.

An example, which we have already used when deriving the critical density, is the monopole degeneracy: any circularly symmetric redistribution of mass inwards of all observed images, and any circularly symmetric change in mass outside all observed images will change τ by at most an irrelevant constant in the image region, and hence have no effect on lensing observables. This means in particular that doubles and quads contain no information about the monopole part of the interior mass distribution, though they constrain the total mass enclosed. So in the example in Sect. 1.1.2 our choice of core radius was irrelevant; it specified the location of the inner critical curve and the outer caustic, but those played no part since images and sources never went near them.

In addition to degeneracies of the above type, which all involve localized changes in the arrival-time surface, there is one special degeneracy which is particularly serious: the mass disk degeneracy [32,94,123,113]. In this the τ scale of the whole arrival-time surface gets stretched or shrunk. To derive it we rewrite (1.11) first discarding a $\frac{1}{2}\vec{\beta}^2$ term since it is constant over the arrival-time surface, and then using $\nabla^2 \vec{\theta}^2 = 4$, to get

$$\tau(\vec{\theta}) = 2\nabla^{-2}(1 - \kappa) - \vec{\theta} \cdot \vec{\beta}. \qquad (1.32)$$

Now the transformation

$$1 - \kappa \to s(1 - \kappa), \qquad \vec{\beta} \to s\vec{\beta}. \qquad (1.33)$$

where s is a constant which just rescales time delays while keeping the image structure the same; but since the source plane is rescaled by s all magnifications are scaled by $1/s$, leaving relative magnifications unchanged. The effect on the lens is to make it more like or less like a disk with $\kappa = 1$. Figure 1.12 illustrates. Note that in (1.33) s can become arbitrarily small; it can not become arbitrarily large because otherwise κ will become negative somewhere in the image region. (Negative κ *outside* the image region can always be avoided by adding an external monopole).

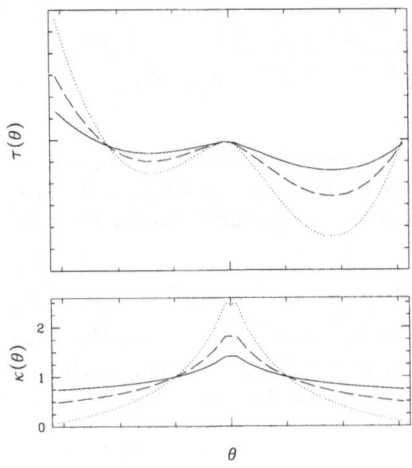

Fig. 1.12. Illustration of the mass disk degeneracy, showing the surface density (lower panel) and the arrival time (upper panel) for three circular lenses. The units, except for κ, are arbitrary. The arrival time indicates a saddle point (which looks like a local minimum in this cut), a maximum, and a minimum. The dashed curves correspond to a non-singular isothermal lens. Stretching the time scale amounts to making lens profile steeper (dotted curves) and shrinking the time scale amounts to making the lens profile shallower (solid curves).

From the modelers point of view, the mass disk degeneracy is a degeneracy in the central concentration of the lens, or the steepness of the radial profile, and we will meet this many-headed monster again in the modeling section. An easy way of remembering its effect is "the lens gets steeper as the universe gets smaller". The lensing data stay exactly the same, and the mass inside an Einstein radius is unaffected [113], but the sources before lensing get larger and brighter, and h gets bigger. Which reminds us that this degeneracy is particularly inimical to measuring h from lensed quasars, where it dominates the uncertainty. In principle it could be broken in various ways: if the intrinsic brightness of sources were known, or if sources at very different redshifts were lensed by the same lens [2], or indeed if h were known from some other method. But there seem no immediate prospects for any of these.

Another kind of degeneracy is associated with a non-lensing observable that is often observed in connection with lensing, velocity dispersion. Lenses follow an approximate relation between Einstein radius (or some surrogate for it in non-circular lenses such as the size of the outer critical curve) and the line-of-sight velocity dispersion:

$$\theta_E \simeq 2'' \times \frac{\langle v_{los}^2 \rangle}{(300 \, \text{km s}^{-1})^2}. \tag{1.34}$$

To see why there should be such a relation, we rewrite the expression (1.6) for the Einstein radius of a circular mass distribution as

$$\frac{GM}{\theta_E D_L} = \frac{c^2}{4} \frac{D_S}{D_{LS}} \theta_E. \tag{1.35}$$

Now, the left hand side in (1.35) will be of order $\langle v_{los}^2 \rangle$ because of the virial theorem, leading to (1.34). The trouble is that the relation (1.34) cannot be made more precise, because the exact coefficient that would go into it depends on the mass distribution in a very complicated way. In general, more centrally concentrated mass distributions would give larger velocity dispersions. On the other hand, an isothermal sphere in stellar dynamics gives $(3\pi/2)\langle v_{los}^2 \rangle$ for the left hand side in (1.35) while a barely compact homogeneous sphere gives $5\langle v_{los}^2 \rangle$ — almost the same number despite the very different mass profile.

1.2 Observations

1.2.1 Historical Background

While the concept of light deflection by massive bodies was already proposed by Isaac Newton in the 18th century [90], the astrophysical and cosmological potential of the phenomenon was, with notable exceptions, taken seriously only after discovery in of the first multiply imaged quasar by Walsh, Carswell & Weymann [133]. The observation of two well separated images of the same source at $z = 1.41$ not only confirmed the existence of what had previously been seen largely as a theoretical curiosity, but also established gravitational lensing as a new field of astrophysics. Indeed, the existence of even a single lensed quasar, lent considerable hope to the application of Refsdal's method [104,105] for determining the Hubble parameter H_0. Proposed in 1964, the method is based on the measurement of the light variations in the lensed images of a distant source. The time lag, or so-called "time-delay" between the arrival times of the luminous signal from each image of the source to the observer, is directly related to

H_0 and to the mass distribution in the lensing object. Measuring the time-delay therefore provides us, via a mass model for the lensing galaxy, with an estimate of H_0. Refsdal originally proposed to apply his method to distant supernovae. The discovery of quasars by Schmidt [121] offered new prospects in using even more distant light sources.

Measuring time-delays is far from trivial: the angular separations between the lensed images are usually small, typically 1-2 arcsec, and not all quasars are willing to show measurable photometric variations. In addition, characterizing the mass distribution responsible for the lensing effect, assuming the lensing galaxy is detected at all, was very challenging at the time of the first discoveries. CCD detectors, were only just coming into use. They were hard to obtain and had small formats and high read noise. The uncontrolled thermal environments of telescopes produced mediocre seeing, typically larger than the angular separations observed in most presently known objects (see Tables at the end of this chapter). The Hubble Space Telescope (HST) was more than a decade off in the future. Despite these difficulties, searching for new systems suitable for cosmological investigations became a major activity in the early eighties. Based on the argument that some of the brightest quasars might be magnified versions of a lower luminosity object (e.g., Sanitt [115]), systematic searches for new multiply imaged sources were undertaken among the apparently brightest quasars. These searches yielded the discovery of more doubles, like UM 673 [126], but also new image configurations. PG 1115+080 [138] was thought to be triple but turned out to be an off-axis quadruple with higher resolution observations [44]. More symmetric quadruples, such as the "cloverleaf" [77], were also found. Almost simultaneously, radio searches yielded their first results. As the radio emitting regions of quasars are larger than the optical ones, lensed radio loud quasars were often found to be complete Einstein rings: MG 1131+0416, MG 1654+1346, PKS 1830-211 [45,66,125]. The observation of complete or partial rings offers more constraints than 2 or even 4 point source images and led to the development of more accurate models [59].

During this same period, systematic campaigns were initiated to measure time delays, much of it concentrated on the first lens discovered: Q 0957+561. Early reports [119], [38] gave contradictory results. Vanderriest et al. [132] and Schild et al. [120] derived a value of 415 days, from ground based optical observations. Press et al. [101] reanalyzed Vanderriest's data and published a very different time delay: 536 days, a value supported by the radio monitoring results obtained at the Very Large Array (VLA) [108]. The dictum attributed to Rutherford [8], "If your experiment needs statistics, you ought to have done a better experiment," appears to have been borne out. Improved optical [65,92] and radio monitorings [42] have finally settled the issue. They reconcile the optical and radio time delays and lead to the value of $\Delta t = 417 \pm 3$ days.

The controversy over Q 0957+561 reflects the difficulty of measuring time delays. Quasars do not commonly show very sharp light variations, and their light curves are often corrupted by the erratic photometric variations induced by microlenses (stars) in the lensing galaxy (see Sect. 1.2.3 of the present Chapter). Photometric monitoring over a period considerably longer than the time delay is therefore necessary. Temporal sampling must also be sufficiently frequent to average out short timescale microlensing variations. Microlensing may corrupt quasar light curves but it is of considerable interest for constraining the statistical mass of MACHOs (see Chapter 1) in the lens [136] and the size of the lensed source [134]. With particularly good data obtained over a wide wavelength range, it might even be possible to reconstruct some of the quasar's accretion disk parameters, such as, size, inclination and details of the spectral energy distribution of the accretion disk as a function of distance from the AGN's center [3,150].

Progress with CCD detectors, with radio interferometers and with image processing techniques has made it possible to overcome at least some of the observational limitations on time delay measurements. The list of systems with know time delays is rapidly growing, with optical and radio time delays both available in some cases. Schechter et al. [116] obtained optical light curves for PG 1115+080 and two time delays between two images and the group of blended bright images A1+A2. Three time delays have been measured from radio VLA observation, in the quadruply imaged quasar CLASS B1608+656 [34]. The two bright radio doubles PKS 1830-211 and B 0218+357 are two other cases with known time delays (e.g., [76,14]). Note also the lucky case of B 1600+434 which has both optical [20] and radio [64] time delays and even overlapping light curves. Many more time delays have recently been obtained at the Nordic Optical Telescope or at ESO [22,23,48].

The level of interest in lensed quasars has followed a more or less predictable course. The considerable excitement following what was effectively the birth of the field in 1979 was followed by extraordinary growth, as measured by the number of papers published [128] and number of lenses known [61]. The phenomenon is no longer be so novel, it is entering a more mature, and astrophysically and cosmologically, more productive stage. Observational and theoretical advances have proceeded in parallel, with considerable improvement in "best" estimates of H_0, in weighing distant galaxies, and in probing their stellar and dust content. And as with other areas of astrophysics, there is an increasing tendency toward large, international teams, marking the substantial demands of the enterprise.

1.2.2 Observational Constraints in Quasar Lensing

Given the small deflection angles involved in multiply imaged quasars, high resolution observations are required to measure accurately the main observables: the position of the quasar images and of the lens, the time-delay, and the magnification at the position of the images.

The Image Configuration and the Time Delay: For most known lens systems it has become relatively straightforward nowadays to obtain astrometry of the requisite precision, especially when HST images are available. However, adequate temporal sampling is also required as soon as the goal is to measure the time delay. Photometric monitoring with 0.1″ resolution would be possible with

HST but one could not realistically expect the large numbers of orbits necessary. Until recently, such work was restricted to radio wavelengths, less affected by weather conditions than optical ones and providing data with higher resolution on a more regular basis. Scheduling is also facilitated as one can observe in day time. But as only 10% of quasars are radio loud, this restricts the available sample of lenses.

Recent advances in image processing techniques have extended the range of ground-based imaging into the subarcsecond regime. Typically reconstruction and deconvolution techniques can only be used with relatively high signal-to-noise observations, but in such cases improvements of a factor of two in resolution have been possible [79].

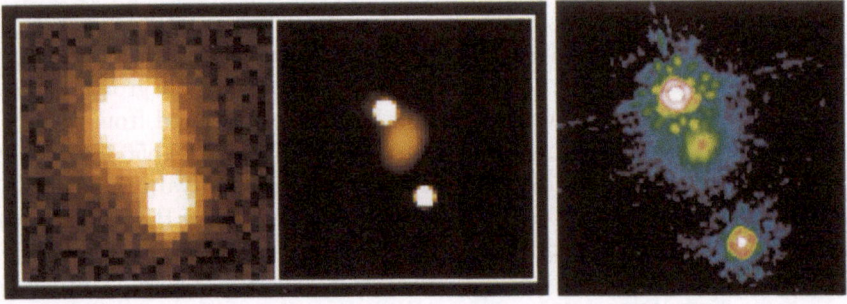

Fig. 1.13. Two ways of obtaining high resolution images of lensed quasars. From left to right: near-IR ground based image of HE 1104-1805. It has been obtained with the ESO/MPI 2.2m telescope in the J-band under average seeing conditions (0.7″). Its resolution has been improved down to 0.27″ on the deconvolved version of the data displayed in the middle panel. The lensing galaxy is obvious, between the QSO images. Its position and elongated shape oriented with a PA of about 30 degree are confirmed by the H-band HST/NICMOS image shown on the right (HST image from the CASTLE survey) (see Color Plate).

Figure 1.13 shows an example of high resolution data of the doubly imaged quasar HE 1104-1805 [140], as might be obtained either with the HST (here, in the near-IR) or from post-processed (deconvolved) ground based images. The data presented in this figure are sufficient to infer the image and lens positions with an accuracy from a few milli-arcsec (quasar images) to a few tens of milli-arcsec (lensing galaxy). In fact, the combination of space observatory data [72] and post-processed ground based data now allow for accurate photometric monitoring in the optical and for detailed modeling.

The range in properties of lens systems is such that there is no single factor which consistently limits one's ability to carry out a determination of H_0 through time delay measurement. It seems that there is no "golden lens," no ideal case that will give a "best" measurement of H_0. In some cases, the error on the time delay dominates (for example HE 2149-2745 [22]), while in other systems more symmetric about the center of the lensing galaxy, the errors introduced by the astrometry of the quasar images will dominate [116,56,28,50]. In still other cases the erratic variations of the light curves introduced by microlensing events in the lensing galaxy are the main source of error [142,20]. It therefore seems more reasonable to monitor as many systems as possible rather than trying to concentrate on a particular one which might have its own unknown sources of systematic errors.

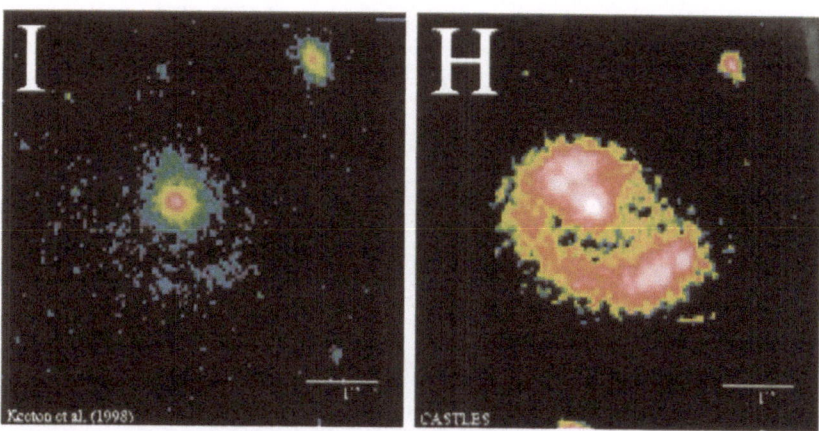

Fig. 1.14. The lensed radio source MG 1131+0456 is a nice example of system with the source only visible in the near-IR and longer wavelengths. On these HST images obtained by the CASTLEs group, only the lens is visible (left) in the optical I-band. On the H-band image (right), the source is seen as a almost full Einstein ring (see Color Plate).

Distances to the Source and Lens. As seen from (1.3), modeling lensed quasars requires knowledge of the distance D_L to the lens, and of the distance D_S to the source. While the lensing galaxies are not especially faint by current standards, measuring their redshifts is non-trivial. In the optical, the background quasar is often bright and hides the much fainter lensing galaxy. In some lucky cases, the lensing galaxy shows emission lines in superposition on the quasar spectrum [130], but this is not the rule. To date, no HST spectrum has been taken of a lensing galaxy, but application of deconvolution techniques [29] to spectra obtained on ground based 10m class telescopes have proved useful and have yielded the measurement of several lens redshifts [75,22].

There are a number of cases where the lens and source have such different spectral energy distributions that they must be observed at very different wavelengths. MG 1131+0456 (see Fig. 1.14) and PKS 1830-211 are examples of systems where the source can be seen only in the near-IR. In the case of PKS 1830-211, the source's redshift could be determined only from IR spectroscopy [74]. Other more extreme cases like MG 1549+3047 show the lens only in the optical/near-IR and the source only in the radio. As such systems show no

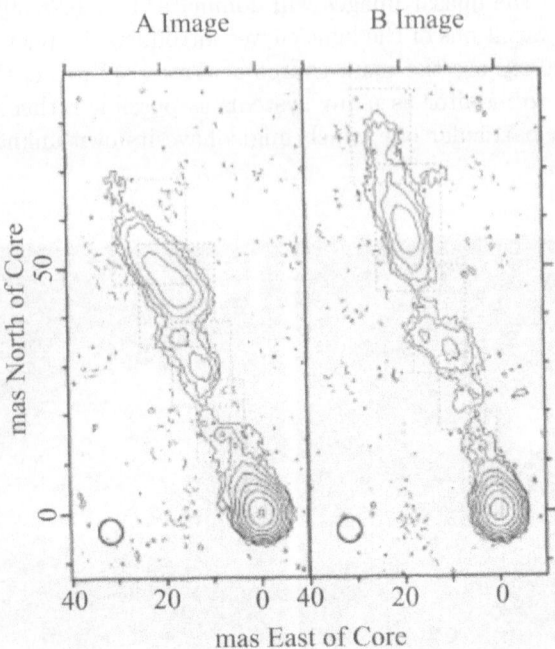

Fig. 1.15. The double quasar Q 0957+0561 observed in VLBI [24] at 6 cm with a resolution of 6 milli-arcsec. The two quasar images show a very detailed radio jet that is used to place constraints on the lens model.

light contamination by the background source, they allow for a detailed study of the lens. In the case of MG 1549+3047, the velocity dispersion of lensing galaxy could even be measured [69,70]. A major drawback however, is that the redshift of the source remains unknown.

The Quasar Host Galaxy and Background Objects. At very high angular resolution, it becomes possible, beyond measuring the position and brightness of the quasar, to resolve details in the distorted and amplified quasar host. Observing the distorted quasar host galaxy brings extra constraints on the lensing potential, and helps to see distant quasar host galaxies (up to redshift 4.5) that would have been missed without the lensing magnification [72,57]. Figure 1.14 shows an example of a red quasar host galaxy where small details are unveiled, at the resolution of the HST (about 0.15″ in the H-band). Such information is of importance as each detail might be identified in the counter-image and used to place additional constraint on the reconstruction of the lensing galaxy's mass profile. In the radio, using Very Long Baseline Interferometry (VLBI) with resolution on the order of the milli-arcsec, "blobs" can be seen in the lensed images of the radio jet in the source. Such observations, producing the spectacular

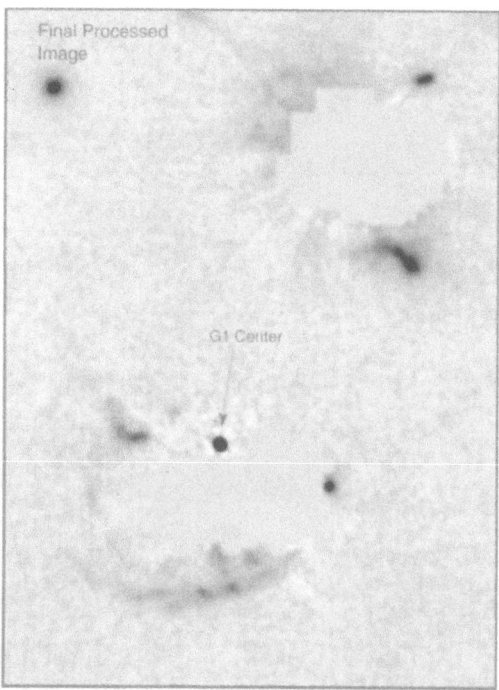

Fig. 1.16. Using the HST, arcs and arclets are discovered in the field of Q 0957+0561 [12] and help to determine the mass profile of the lensing galaxy. The quasar components have been subtracted on this STIS image, provided to us by Gary Bernstein and Phil Fischer, prior to publication. Several arcs are visible as well as members of a foreground galaxy cluster at z~0.35 [5]. G1 is the center of the main lensing galaxy which has also been removed from the image.

maps shown in Fig. 1.15 are restricted to very few objects with such high spatial resolution [24,110,131].

Arcs and arclets are sometimes seen in the immediate vicinity of the quasar components. These objects, as shown in Fig. 1.16, might be companions to the quasar or simply unrelated background sources [12]. As in the case of the quasar host, they probe the lensing potential, with the further advantage that they do no lie at the same position as the quasar and therefore probe the lensing potential in a location otherwise inaccessible.

Intervening Clusters/Groups. Isolated lenses may be the exception rather than the rule. Multiply imaged quasars often lie close to the line of sight to foreground groups and even clusters of galaxies. A massive galaxy cluster ($\sigma \sim 600 - 1000\,\mathrm{km\,s^{-1}}$), even situated several tens of arcsecs away from a system, will modify the expected image position and the time delay, hence also modifying the infered value for H_0. Therefore one has to set constraints not only on the astrometry and shape of the main lens, but also on additional objects

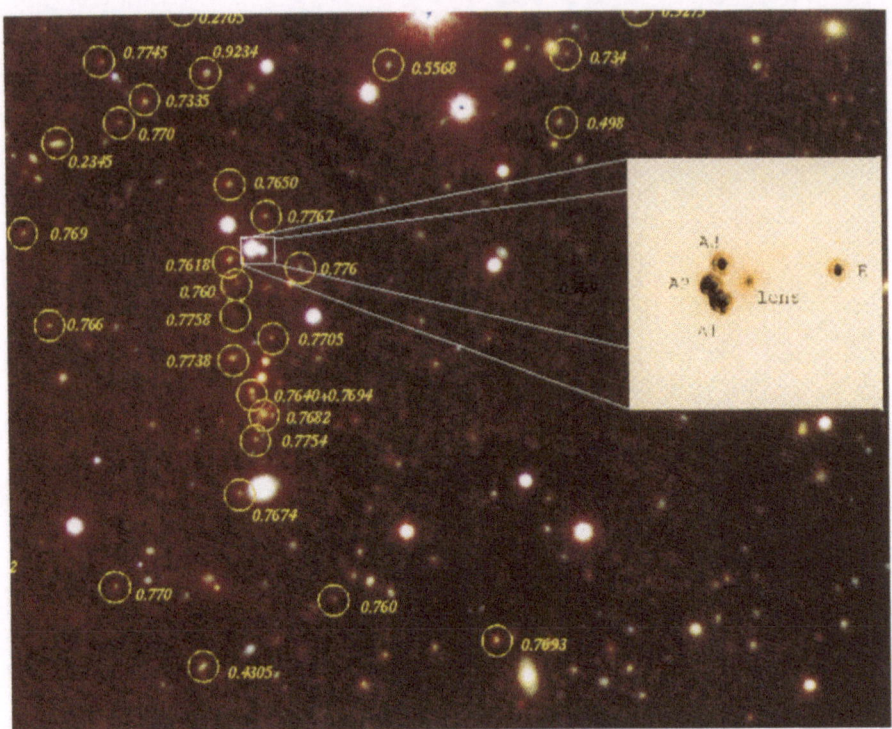

Fig. 1.17. The quadruply imaged quasar RX J0911+0551 [7] at $z = 2.8$ and the intervening cluster at $z = 0.77$ [58] which significantly modifies the overall gravitational potential responsible for the lensing effect. The field of view is $3.5' \times 3.5'$ (see Color Plate).

that may modify the total gravitational potential responsible for a given image configuration.

In the case of B 1600+43 [54,20] the lens is an edge-on spiral at $z = 0.41$ [33] with a lower redshift spiral a few arcsec South-East [54]. The quadruply imaged quasar PG 1115+080 can be modeled only by taking into account a small group of 4-5 galaxies at $z = 0.31$ (the same as the lens) about $20''$ away from the line of sight [56]. Two spectacular examples of intervening clusters are RX J0921+4529 which is situated in an X-ray cluster at $z = 0.32$ [87] and RX J0911+0551 (see Fig.1.17). The later is lensed by a galaxy at $z = 0.77$, a member of an X-ray cluster centered about $30''$ to the south [19,86]. The cluster's velocity dispersion has been measured from the redshifts of 24 members [58] as 836^{+180}_{-200} km s^{-1}

1.2.3   Microlensing of the Quasar Images

As was shown in Sect. 1.1.3 that microlensing of individual quasar images is not unexpected, although it depends upon the size of the source and the constituents of the lensing galaxy. The angular scale associated with such microlenses is smaller than that of the macro lens by $\sqrt{(M_{\mathrm{micro}}/M_{\mathrm{macro}})}$. Stellar mass microlenses therefore produce microarcsecond splittings, not accessible to present-day instrumentation. But the image magnification by microlenses is nonetheless observable – flickering of the combined flux from the unresolved microimages can be detected as the stars are move randomly in the lensing galaxy. While the observational evidence for microlensing is fragmentary, there is enough to indicate that this phenomenon, predicted immediately after the discovery of Q 0957+561 [25], plays an important role in the lensing of quasars.

The first hints of microlensing were found in the doubly imaged quasar Q 0957+561 [132]. The difference light curve between the two components (once corrected for the time delay) showed slow variations unrelated to the intrinsic variability of the quasar. These additional variations are thought to be the explanation of different time delays measured by different investigators. They have also been identified as a potentially interesting tool to set constraints on the stellar content of the lens and on the internal structure of the lensed quasar on parsec scales (see for example [103,96,134]).

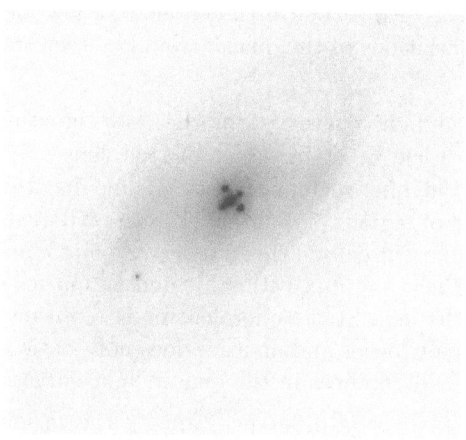

Fig. 1.18. HST V-band image of Q 2237+0305. Four quasar images at $z = 1.69$ are seen about $1''$ away from the nucleus of a much lower redshift lensing galaxy ($z = 0.04$). The high density of stars in the lens' nucleus and their high projected angular velocity make of Q 2237+0305 a privileged object for the study of microlensing.

Q 2237+0305, also known as the "Einstein cross" (see Fig. 1.18), was quickly recognized after its discovery [49] to be particularly susceptible to microlensing. The redshift of the lens, $z_L = 0.04$, is so low that the apparent angular velocity of the microlenses, in projection on the plane of the sky, is much higher than in other systems. Moreover the Einstein rings of thes microlenses have a larger angular diameter, making it more likely that they are larger than the source. The quasar is therefore expected to show frequent and rapid variations, the mean time separating each microlensing event being approximately the time required for the microlenses to run across a distance equal to the diameter of their Einstein ring (see (1.7)). Note however that time scales involved can be significantly different from those calculated with this naive approach. As can be seen from Fig. 1.10 there are regions of high magnification, in particular close to the cusps of caustics, which are exceedingly narrow, much smaller than the projection back onto the source plane of the Einstein ring [134]. There are also plateaus, larger than the Einstein ring, over which the magnification is relatively constant, and usually less than unity.

As the optical path to each quasar image intersects the lensing galaxy at very different locations, microlensing-induced variations in the light curves of the quasar images are uncorrelated. Intrinsic variations of the quasar would be seen identical in each image, separated by the time delay. Time scales involved for microlensing events in the Einstein cross were predicted to be of the order of a few months [134], spectacularly confirmed by optical monitoring [93,149]. This is much longer than the time delay of the system (about 1 day), making it easy to discriminate between intrinsic variations of the source and microlensing events. Figure 1.19 illustrates this: erratic variations of the 4 light curves (especially component C) are seen, with a typical time-scale of a few months. At the scale of the plot, intrinsic variations of the quasar would be seen simultaneously in all light curves.

Flickering of quasar light curves is not the only signature of microlensing. As noted in Sect. 1.1.3 the Einstein radius of microlenses is small and may be comparable in size to the inner regions of quasars. One may therefore observe differential magnification of regions of different sizes. As different regions of quasars are thought to have different colors, this implies *chromatic* magnification. There are many instances where the flux ratios for quasar images are quite different at different wavelengths [85]. Static microlensing is often invoked as a possible explanation. If one region varies and another does not, static microlensing might also produce chromatic differences in the quasar light curves.

More generally, the spectral differences among the regions of a quasar will involve the presence or absence of emission lines. One might therefore expect differential magnification of the emission lines and continuum. The deblending of closely separated quasar spectra is not trivial. Fortunately, the relatively wide angular separation system, HE 1104-1805, appears to show the phenomenon [140,143]. In this double system (see Fig. 1.13), the spectra of the two components are identical in the emission lines but show a different continuum, suggestive of microlensing. Figure 1.20 shows the difference spectrum (bottom panel) between the two quasar images. In order to subtract properly the emission lines from the spectrum of component A (top panel), one has to subtract a scaled version of the spectrum of component B (middle panel). The scaling factor of 2.8 is found to be stable with time and wavelength, even in the near-IR [30]. This suggests that emission lines are unaffected by microlensing. The difference continuum is blue and shows photometric variations. Part of these variations are intrinsic to the quasar, and are used to infer the time delay of the system, but additional

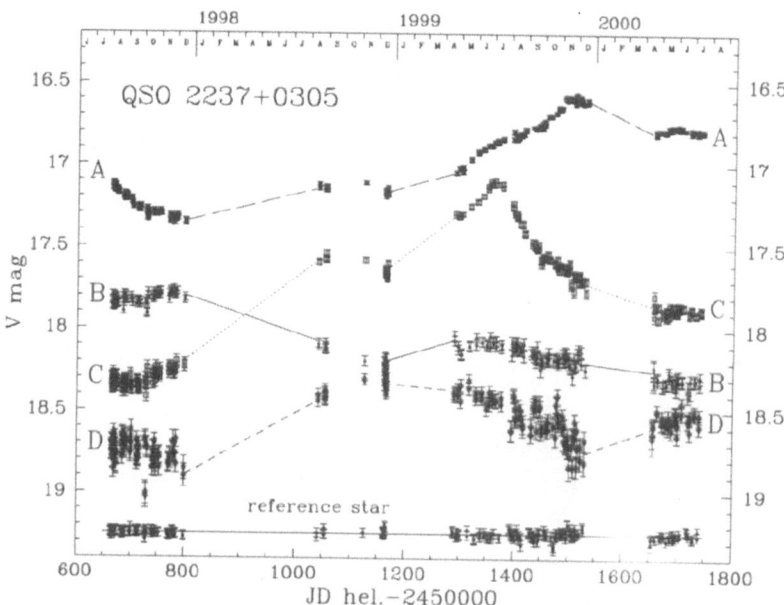

Fig. 1.19. Optical V-band light curves for the four quasar images in the Einstein cross, Q 2237+0305. The time-delay in this system is of the order of a day. The very different behaviour of the four light curves, with slow variations of the order of a month, strongly support the idea of microlensing induced variability [149].

flickering can be attributed to microlensing [143]. With higher signal-to-noise spectra of HE 1104-1805, it has been found that some emission lines might be affected by microlensing as well [75]. For example the red side of the CIII emission line does not subtract perfectly after subtraction of the B spectrum. This is also true for HE 2149-2745 [22], but there are no similar observations so far for other systems.

In principle, this effect can be used to study the detailed structure of active galactic nuclei. For the sake of argument, one adopts a standard model for of active galactic nuclei [17] where, for example, the continuum region is much smaller than the broad line region. The continuum region itself is composed of an accretion disk which radiation field is less and less energetic when going from the center of the accretion disk to its outer parts. As wavelength decreases with increasing energy, the accretion disk radiates bluer photons in the center than in the outer parts. Let us now make the realistic assumption [134] that the mean size of the micro-caustics corresponds about to the angular size of a quasar accretion disk. Simple geometric considerations shows that the smaller regions (compared with the caustics) are entirely magnified, while only a fraction of the larger regions are amplified. That is, the emission line region will be less magnified by microlensing than the continuum, and the outer parts of the accretion disk will be less magnified than the inner parts. In other words, the flickering of the light

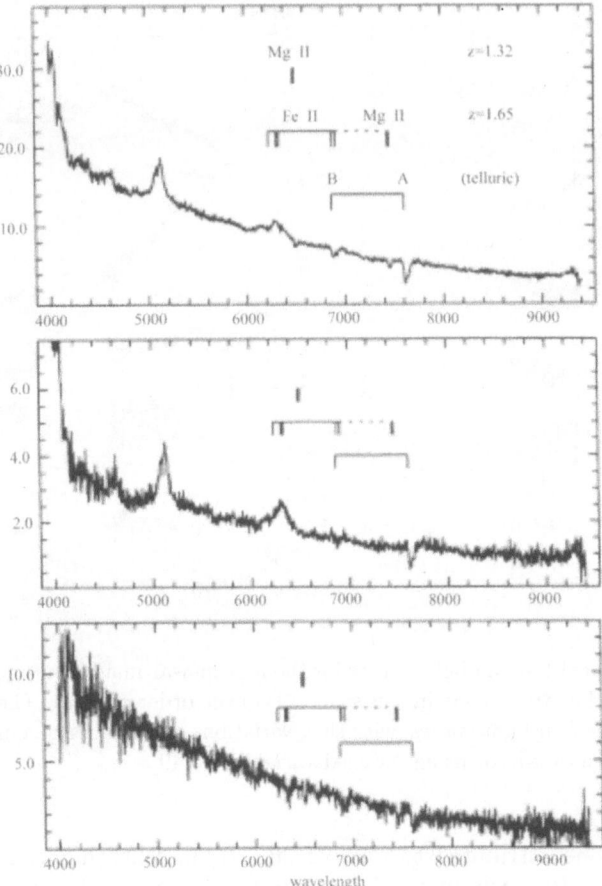

Fig. 1.20. Spectra of the two quasar images in HE 1104-1805 [140]. The two first panels starting from the top show the quasar spectra. The bottom panel shows the difference between the spectrum of the brightest component of the system and a scaled version of the spectrum of the faint component. A scaling factor of 2.8 is necessary to subtract the emission lines from the spectrum. The labels in the figure are related to absorption lines by the lensing galaxy.

curves expected from microlensing should be stronger in the blue than in the red, and should even be invisible in the emission lines.

This very simple scheme is a lot more complicated in practice, simply because one can not map the actual network of micro-caustics present in a given lensed system: this would require a map of the mass distribution in the lensing galaxy ! Still, one can propose a quasar model and predict the statistical behaviour of the light curves, as a function of wavelength. Such a theoretical work has been investigated [3,150,151], with the goal to derive the relative sizes of quasar emission regions. Figure 1.21 illustrates how a given distribution of microlenses

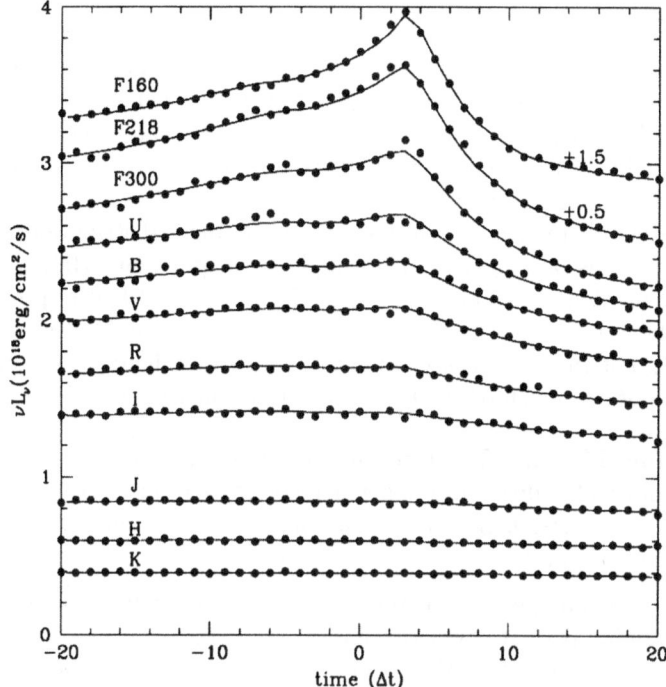

Fig. 1.21. Expected light curves for the microlensing events in a quasar at $z = 1.695$ [3]. Microlensing events have larger amplitudes (larger magnification) at short wavelengths. The time axis is in arbitrary units which depends on the velocity and redshift of the microlenses.

preferentially magnifies the innermost (blue) regions of quasar accretion disks, hence producing light curves with luminosity peaks progressively increasing while observing from the near-IR K-band to the ultraviolet U-band [3].

Unfortunately, the amount of data available so far is too small to implement any of the proposed method to probe quasar structure from microlensing.

1.3   Models

The small number of observables in lensing means that the observational data, no matter how accurate, can be fit by a huge variety of lens models. The space of allowed models must be narrowed by the adoption of priors which reflect our understanding of the relative astrophysical plausibility of different mass models. There are two strategies for doing this. One is to adopt a parameterized mass distribution, where the parameters are chosen to include the reasonable and important variations expected among lensing galaxies. The other strategy is to keep the mass map free-form, but impose astrophysical priors as constraints on it.

We now discuss both these modeling strategies.

1.3.1 Parameterized Models

When building parametrized models, the small number of observables then demands a small number of parameters. Eliminating a parameter (e.g. the octupole moment of the gravitational potential) means that some aspect of the lensing galaxy is not being modeled. A wise choice of parameters models those aspects which are important for the task at hand.

Fortunately there is a vast literature on the mass distributions and gravitational potentials of galaxies. For example (and quite importantly for the interpreting time delays) we know that galaxies have mass density profiles which vary roughly as $1/r^2$, giving galaxies flat rotation curves and flat velocity dispersion profiles. For the sake of discussion we put forward here a "standard" model which incorporates much of what we know about nearby galaxies.

Some Simple Models. We start with the simple monopole potential and describe a number of additional terms which correspond, at least roughly, to what one might expect for galaxies in a variety of contexts, adding degrees of freedom which we have reason to believe nature exploits.

Singular isothermal sphere The singular isothermal sphere is a cornerstone of galaxy dynamics [15]. It gives the flat rotation curves and constant velocity dispersion profiles characteristic (to a first approximation) of spiral and elliptical galaxies respectively. It has a three dimensional potential $\Phi = v_c^2 \ln r$, where the circular velocity $v_c = \sqrt{2}\sigma$, with σ the one dimensional velocity dispersion. Integrating this along the line of sight and multiplying by $2D_{LS}/(D_L D_S c^2)$ gives gives the lens potential

$$\psi(\vec{\theta}) = \theta_E \theta, \qquad (1.36)$$

which is the same as (1.27) but now with with $\theta_E = 4\pi\sigma^2 D_{LS}/D_S c^2$ (measured in radians) giving the lens strength. We recall (cf. (1.28)) that such a lens produces two colinear images, one on each side of the lens, with magnifications given by (1.29). The infinite second derivative of the potential at the origin gives infinite demagnification of a central third image.

Power-law monopole The singular isothermal sphere is a special case of the power-law monopole

$$\psi(\vec{\theta}) = \frac{\theta_E^2}{(1+\alpha)} \left(\frac{\theta}{\theta_E}\right)^{1+\alpha}, \qquad (1.37)$$

where the central concentration index, α, measures the deviation from isothermality. The normalization has been chosen so that the strength θ_E is again the radius of the Einstein ring. As the exponent α approaches -1, the potential approaches that of a point mass.

Self-similar power law quadrupole Since galaxies are not circularly symmetric, there is no reason why their effective potentials should be. The flattening of a potential is dominated by a quadrupole term, which if we use polar coordinates with $\theta = (\theta, \phi)$, varies as $\cos 2\phi$. A simple model for the effective potential which incorporates the non-negligible quadrupole of galaxies incorporates quadrupole term with the same radial dependence of the monopole, giving equipotentials which similar scaled versions of each other,

$$\psi(\vec{\theta}) = \frac{\theta_{\mathrm{E}}^2}{(1+\alpha)} \left(\frac{\theta}{\theta_{\mathrm{E}}} \right)^{1+\alpha} [1 + \gamma \cos 2(\phi - \phi_\gamma)]. \qquad (1.38)$$

An on-axis source gives 4 images whose distance from the lens center is approximately equal to the lens strength θ_{E}. Dimensionless γ gives the flattening of that quadrupole and ϕ_γ gives its orientation.[4] The special case $\alpha = 0$ gives a flattened system with the flat rotation curve and constant velocity dispersion profile characteristic of isothermals. While the equipotentials are self similar for all α, the equipotentials and the equidensity contours are both self-similar only for the $\alpha = 0$, isothermal case.

Tidal quadrupole (first order tide) Equipotentials which have the same shape are esthetically appealing but highly idealized. In particular galaxy equipotentials will deviate strongly from self-similarity if the quadrupole is due to a tide from a neighboring galaxy or cluster of galaxies. In that case the quadrupole term shows a θ^2 dependence on distance from the center of the lens, as was seen in (1.30). Among others, [65] have noted that the quadrupoles of many lensed systems appear to be due to tides rather than to the flattening of the lensing galaxies. A simple potential incorporating these features is

$$\psi(\vec{\theta}) = \frac{\theta_{\mathrm{E}}^2}{(1+\alpha)} \left(\frac{\theta}{\theta_{\mathrm{E}}} \right)^{1+\alpha} + \tfrac{1}{2}\gamma\theta^2 \cos 2(\phi - \phi_\gamma). \qquad (1.39)$$

This is much like the self-similar power law quadrupole of (1.38). While the monopole term is, as in the previous cases, a power law, the quadrupole term has the θ^2 dependence characteristic of a tide. In general we expect a lens to have both a tidal quadrupole, from neighboring galaxies, and something like the self-similar quadrupole due to the flattening of the lensing galaxy itself.

Clusters as mass sheets Galaxies typically reside in groups and clusters, with considerably more mass (dark matter) associated with the cluster than with the individual galaxies. One must therefore take the gravitational potential of the associated cluster into account. The scale of a cluster is much larger than that of a galaxy, so its surface density projected onto the galaxy is to first order constant. A mass sheet of uniform density produces an effective potential

$$\psi_s(\vec{\theta}) = \tfrac{1}{2}\kappa_s\theta^2 \qquad (1.40)$$

where κ_s is the dimensionless convergence associated with the mass sheet.

As was shown in Sect. 1.1, differentiating twice one finds that a superposed mass sheet stretches an image configuration by a constant factor $1/(1 - \kappa_s)$

---

[4] Following Kochanek (1992) we (somewhat confusingly) use the same symbol, γ, for the flattening as is used for the shear. The flattening and the shear are equal at $\theta = \theta_{\mathrm{E}}$, but not elsewhere.

without changing any of the dimensionless ratios associated with the image configuration. A model that failed to take account of such a mass sheet would predict too long a differential time delay by the just this same stretch factor. But there is no way of knowing from image positions or relative magnifications whether or not such a mass sheet is present. This formal degeneracy *demands* that one bring to bear "external" information regarding the projected density of any such mass sheet.

Clusters and higher order tidal terms In the above paragraphs we have identified two distinct effects of clusters of galaxies: they introduce tidal and mass sheet terms into the effective lensing potential. There are many lenses for which the first order tidal terms are so strong (e.g. [65]) that higher order terms are likely to be important. The simplest way to do this is to drop the tidal term above and to model the cluster as a isothermal at position $\vec{\Theta}$ with effective potential

$$\psi_c(\vec{\theta}) = \Theta_E |\vec{\Theta} - \vec{\theta}|. \tag{1.41}$$

This model has three free parameters (replacing the two tidal parameters, γ and ϕ_γ), the lens strength $\Theta_E = 4\pi\sigma^2 D_{LS}/D_S c^2$ where σ is the the velocity dispersion of the cluster, and the polar coordinates (Θ, Φ).

While the cluster potential can be written quite compactly in this form, it obscures the connection between the cluster properties and the tidal and mass sheet terms described above. Taking the lensing galaxy to be at the origin of our coordinate system, we can expand the cluster potential in powers of θ/Θ, where Θ is the distance of the cluster from the origin. Dropping constant terms, we find

$$\psi_c(\vec{\theta}) = -\Theta_E \vec{\Theta} \cdot \left(\frac{\vec{\theta}}{\Theta}\right)$$

$$+ \frac{1}{4}\Theta_E \Theta \left(\frac{\theta}{\Theta}\right)^2 - \frac{1}{4}\Theta_E \Theta \left(\frac{\theta}{\Theta}\right)^2 \cos 2(\phi - \Phi)$$

$$+ \text{ terms of order } \left(\frac{\theta}{\Theta}\right)^3 \text{ and higher.} \tag{1.42}$$

The first term gives a constant deflection Θ_E away from the cluster, showing that the source position $\vec{\beta}$ may be rather far from the origin and the lensing galaxy. The second term is just that of a mass sheet with $\kappa_s = \frac{1}{2}(\Theta_E/\Theta)$ while the third is a tidal term with shear $\gamma = \kappa$. Noting that the coefficient of the shear term is negative, we find that the position angle of the shear, ϕ_γ, as defined in (1.38) and (1.39), is at right angles to the position angle of the cluster, Φ.

The equality of the shear and convergence suggests a possible resolution of the mass sheet degeneracy: measure the shear and infer the convergence. We adopt this approach with an obvious caveat. To the extent that clusters and groups are *not* isothermal, such a "shear inferred" mass sheet correction will introduce a systematic error in a derived Hubble constant.

The terms of order $(\theta/\Theta)^3$ are useful because they break the classical tidal degeneracy. Since $\gamma = \frac{1}{2}(\Theta_E/\Theta)$, we might produce an equally strong first order tide by putting an isothermal cluster with twice the Einstein radius at twice the distance. Alternatively, we might put the cluster at position $-\vec{\Theta}$ without changing the first order tide. Keeping the higher order terms resolves these ambiguities. But rather than add many coefficients, it is conceptually simpler and more economical to replace the two parameters of a first order tide with the three parameters of a circularly symmetric cluster. There are several lenses (e.g. RX J0911+0551, PG 1115+080 and B1422+231) for which higher order tidal effects have been used to determine the position and lensing strength of the associated cluster.

Yet more degrees of freedom Even in the absence of tides, there is no reason to insist that the monopole and quadrupole terms of a galaxy potential have the same dependence on θ, *i.e.* that the potential be self-similar. The self-similar model presented above can readily be extended to allow for separate θ exponents, permitting the potential to get rounder or flatter with increasing θ. Nor is there any reason, in principle, why we should limit ourselves to monopole and quadrupole terms. Purely elliptical density profiles produce potentials with higher order multipoles. Some ellipticals are "boxy" while others are "disky" [11], and these too should have higher order multipoles. Power laws like (1.37) and (1.38) give unphysical mass and density divergences, and should in principle be cut off at small or large θ or broken somewhere in between. Finally, we might argue that it is naive to assume that the dark matter in a galaxy is centered on its starlight, and that we should take the central coordinates of the lensing potential to be free parameters.

With all these possibilities, it is no surprise that different investigators modeling the same system come up with different potentials and derive different values of H_0 from the same time delay. The number of measurements which constrain the potential is small, so one cannot allow oneself the luxury of adding extra parameters just for the sake of insurance. In introducing new parameters the two questions to be kept in mind are the degree to which they degree to which they affect the deflections and distortions, which constrain the potential, and the degree to which they affect the delays, which give the Hubble constant.

Useful Approximations and Rules of Thumb. For the sake of simplicity, suppose that a lens has the power-law monopole potential of (1.37). Using the lens equation, we substitute the gradient of the effective potential, $\partial\psi/\partial\vec{\theta}$, for the deflection, $\vec{\theta} - \vec{\beta}$, in the time delay equation. Under the assumption that two quasar images, A and B are roughly equidistant from the center of the potential, we predict a differential time delay

$$\tau_B - \tau_A \approx \tfrac{1}{2}T_0\left(\theta_A{}^2 - \theta_B{}^2\right)(1-\alpha), \qquad (1.43)$$

where T_0 is the time scale defined in (1.9) and θ is measured in radians. Had we not assumed circular symmetry, θ_A^2 and θ_B^2 would each have a coefficient which differed from unity by a factor of order gamma, usually less than 10%. Equation (1.43) has the important and useful property that it depends only upon observable quantities, assuming that the position of the lensing galaxy can be measured.

Several useful lessons can be drawn from (1.43). First, the more distant image leads the closer image (cf. Fig. 1.8). Second, if $\theta_A \approx \theta_B$ high astrometric accuracy is needed in measuring the position of the lensing galaxy for high precision in the predicted time delay. Third, the predicted delay scales as the square of the separation. The differential time delay of Q 0957+561, 1ʸ2, is therefore atypically long, resulting from its large (6.1″) separation and relative asymmetry. Fourth, if the lens potential is more sharply peaked than a singular isothermal sphere, the predicted time delay is longer. In particular, a point mass model, with $\alpha = -1$, predicts a time delay twice as large (yielding a Hubble constant twice as large for a given observed delay) as the corresponding singular isothermal, $\alpha = 0$ model. Either α must be measured with high accuracy from the observed image configuration or we must bring external considerations to bear upon our models. In comparing models by different investigators for the same system, one must pay particular attention to the way in which the degree of central concentration has been treated.

Fitting Models. *How and what to fit ?* On first thought it seems straightforward to adopt a lens potential and a source, find the predicted images, compare those with the observed images and adjust the parameters associated with the potential and the source so as to get better agreement. On closer examination one discovers that the lens equation can only rarely be solved in closed form for image positions. Worse yet, one finds that small changes in parameters can cause pairs of predicted images to merge and disappear. What does one do in a gradient search when a small trial step causes an image to cease to exist ? Fortunately robust methods for fitting data have been developed, some of which are publicly available [57].

The fluxes of images can readily be measured to 1% accuracy, but the differences between optical and radio flux ratios are of order 10-30%. Given the very much greater accuracy of positions, one might be tempted to dispense with magnifications entirely. But for double lenses, even the simplest non-circular models have one too many parameters to permit fitting using positions alone. Moreover, fitting fluxes can help avoid aforementioned disappearing image problem. It is therefore helpful to use fluxes, but with full awareness of the associated pitfalls.

Image positions constrain the first derivative of the effective potential. Magnifications constrain the second derivative. In some systems more than one time delay can be measured. The first measured delay goes to solving for the dimensioned combination of angular diameter distances in (1.43) but the ratios of the second and subsequent delays to the first give dimensionless constraints on the effective potential itself. Though not yet incorporated in most parametrized models, such constraints are are in principle quite powerful. But a disadvantage so far is that in practice the uncertainties in all measured delays for a given system are roughly the same, as measured in days; so while the fractional uncertainties in the longest delay is typically better than 10%, the fractional uncertainties in the shorter delays are correspondingly greater.

What Constitutes "Good Enough"? There is little difficulty in finding models for lens systems which fit the observed data perfectly. The number of constraints is small, and the number of free parameters is large, and so it should be possible to find an N parameter model which fits the N available constraints perfectly. But that leaves no room for reality testing. Ideally one hopes to find a model with $< N$ parameters for which the predicted images agree with the observed images within the measured uncertainties, giving an acceptable fit to the data.

The words "unacceptable fit" have a damning ring which tends to end discussion. Were we able to measure the relative positions of the lensed images to one part in a million, the deflections due to individual stars within the the lensing galaxy become important. At that point we would be unlikely to ever get an acceptable fit from a macromodel. But the differences in the time delays induced by such microlensing are small.

For the purpose of interpreting time delay measurements a less stringent definition of acceptable may be in order. Consider the case of Q 0957+561. Errors in the positions of 100 milliarcseconds introduce negligible changes in the time delay predicted by (1.43). While one can concoct a parameterized model for which small differences in the positions produce large changes in the predicted time delays, these are, with the exception of the central concentration degeneracy, somewhat artificial.

The Central Concentration Degeneracy. The central concentration degeneracy has already surfaced in our discussion, first theoretically as the mass disk degeneracy, and then in the approximate rule for computing time delays, equation (1.43), and it appears yet again in connection with free-form models. It has also surfaced many times in the literature. A particularly thorough treatment can be found in [131], though it is evident as well in other works [13,116]. Briefly, it has proven exceedingly difficult to constrain the (radial) second derivative of the monopole term of the effective potential. Several factors contribute. In double systems the associated parameters are coupled to the quadrupole amplitude. In quadruple systems the images all tend to lie at roughly the same distance from the center of the lens – otherwise the system wouldn't be quadruple. The radial displacements of these images depend not only on the concentration parameter but also on higher order multipoles. Einstein rings may be less susceptible to this degeneracy because the rings are resolved in the radial direction, though this is controversial [62,114].

What makes this degeneracy pernicious is its strong influence on the predicted time delay, increasing them by a factor $(1 - \alpha)$ in the parameterization of our power-law models (equations (1.37) and (1.38)). In the face of this, one has two choices: to search for a "golden" lens which doesn't suffer from it or to bring external constraints to bear.

Golden lenses, at least 24 carat golden lenses, are rare. MG J0414+0534 would at first sight seem as good a candidate any, with a core and 3 VLBI features, each quadruply imaged. But [131] conclude "It is clear that useful information on the radial profile of MG J0414+0534 is unavailable from this data." Alas even if it were, the object has shown little sign of variability [83,6].

Measurements of velocity dispersion gradient [130] have been made of the lensing galaxies in Q 0957+561 and PG 1115+080, which in principle constrain the degree of central concentration of the potential. This is a particularly difficult measurement because of the competition from the optical images of the lensed quasar. Moreover the effective radius of the lensing galaxy tends to be considerably smaller than the Einstein ring, making it difficult to obtain measurements out to the region of interest.

An alternative approach [109] is to use what one knows about the potentials of nearby elliptical galaxies. They compiled data on the potentials of nearby elliptical galaxies for which not only velocity dispersions but higher order moments of the line of sight velocity distribution had been measured. Their data show a mean power-law index $\langle \alpha \rangle = -0.2$, with a scatter of roughly 0.2 about that value. This is somewhat more centrally concentrated than for the isothermal index, $\alpha = 0$, but not nearly so concentrated as the point mass index, $\alpha = -1$. If the data fail to constrain the power-law exponent, fixing it at its mean value would introduce errors in the predicted time delays of roughly 20%. But since the observed power-law index is so close to the isothermal value, $\alpha = 0$, and since the power law index makes so little difference in the quality of the fit (otherwise it would be well constrained), one does little harm in fixing the power-law index at its isothermal value and making a post-hoc correction to the predicted time delay.

A Proposed "Standard" Model for Lenses. As the preceding sections make only too clear, predicted time delays and derived Hubble constants depend sensitively upon how lens potentials are modeled. In particular, they are sensitive to the degree of central concentration of the lens model, which is especially difficult to constrain using lensed images alone. Such model differences have led to widely divergent predicted delays and derived Hubble constants for what are essentially the same data.

Absent the discovery of a 24 carat lens, one can still make progress measuring the Hubble constant by accepting that most lenses are underconstrained and adopting a "standard" model for which the associated systematic errors are well understood and which is sufficiently simple that it can be applied to a large fraction of the known lensed systems.

The proposed standard In the belief that it will take us within striking distance of H_0, we propose the following "standard" effective potential,

$$\psi(\vec{\theta}) = \theta_E \theta + \tfrac{1}{2}\gamma\theta^2 \cos 2(\phi - \phi_\gamma), \tag{1.44}$$

which is the isothermal variant of the tidal power-law plus quadrupole of (1.39). To the extent that they are understood, the systematic and random errors associated with this model are as follows.

As noted above the assumption of isothermality, $\alpha = 0$, introduces a systematic error, but this can readily be corrected by multiplying the predicted time delay by the factor $1 - \langle\alpha\rangle$. We choose to fit $\alpha = 0$ because in most cases the availalable data fail to constrain α any better than this external constraint and to avoid fussing about second and third generation standards as the appropriate mean value of α is further refined.

In double systems there are too few constraints to permit discrimination between the tidal isothermal as in our proposed standard model and a self-similar isothermal. Among quadruple systems tides appear to be more important than the flattening of the lenses [56], but then tides may explain the relatively large number of quadruple systems [65].

For our proposed standard the differential time delay is given by

$$\tau_B - \tau_A \approx T_0 \times \{ \ \theta_A{}^2 \ [1 + \gamma \cos 2(\phi_A - \phi_\gamma)]$$
$$-\theta_B{}^2 \ [1 + \gamma \cos 2(\phi_B - \phi_\gamma)]\}. \tag{1.45}$$

Had we instead adopted the isothermal variant of the self-similar power-law potential of (1.38), the square bracketed terms would have reduced to unity as in (1.43). If we have made the wrong choice, and if the orientation of the shear, ϕ_γ, is random, our choice of a tidal quadrupole introduces a random error but not a systematic one. If γ is small, the effect is not large. If $\gamma > 0.1$, the quadrupole term is so large that an external tide seems the more likely possibility. So either we make a small random error or we make the right choice.

If we believe that the shear is largely tidal, it seems reasonable to assume that the tide is due to an isothermal potential, and that there is an associated convergence $\kappa = \gamma$. The predicted time delays of (1.45) should therefore be multiplied by a factor $(1-\gamma)$ to account for the "mass sheet" associated with the tidal perturber. We cannot avoid making a systematic error here, but we make a larger error in failing to correct for the projected surface densities associated with tides than in making the correction. Our doubly corrected prediction is therefore

$$(\tau_B - \tau_A)_c = (1 - \langle\alpha\rangle)(1 - \gamma)(\tau_B - \tau_A). \tag{1.46}$$

In summary, our standard model is a tidal singular isothermal. We fit the model to the available constraints and use (1.45) to compute the time delay. We apply a correction factors of $1 - \langle\alpha\rangle$ and $1/(1-\gamma)$ to the predicted time delay to account for, respectively, the mean power-law index observed in ellipticals and the projected surface density associated with tides.

Application of the proposed standard Our standard model is by no means new. It is one of two models used by the CASTLES group to analyze the lens data they have assembled. We note, however, that they do not apply the corrections for central concentration and convergence that we adopt in the previous section.

The CASTLES model for PG 1115+080 has a shear of 0.12 with a predicted C-B time delay of $12\overset{d}{.}8$ for an $h = 1$ EdS universe. Applying the corrections of (1.46) using a mean concentration, $\langle\alpha\rangle = -0.2$, gives a predicted time delay of $13\overset{d}{.}5$. Using Barkana's value of $25\overset{d}{.}0$ for the observed delay gives $h = 0.53$.

The CASTLES group has not yet posted a SIS+shear model for RX J0911+0551, but Schechter gives a shear of 0.307 and a predicted time delay of $120\overset{d}{.}5$ between B and (A1+A2+A3). Again applying (1.46) we get a corrected prediction of $100\overset{d}{.}2$. [48] report a time-delay measurement of 146^d, giving $h = 0.73$ for an EdS universe.

1.3.2 Free-Form Models

Free-form models build a lens as a superposition of a large number of small components, with minimal assumptions about the form of the full lens. They are motivated by three considerations.

1. The fewness of observables in quasar lensing, and the presence of degeneracies, means that any one lens reconstructed from observations is highly non-unique. One needs some systematic way of searching through possible lens reconstructions.
2. The high accuracy of observations, despite their fewness, means that data always show deviations from the parametrized models discussed above. Models with more parameters can fit the data to observational accuracy, but it is not known what all the essential parameters are. Are twisting isodensity contours important ? Does ellipticity vary significantly with radius ?
3. The most important observational constraints from lensing (being image positions, tensor magnifications, and time delays) are linear, which makes it straightforward to fit lenses by superposition.

We will refer to the small components as pixels, but in fact they can be any kind of components and not necessarily small. For example, they may be Fourier or harmonic terms in the mass profile [131]. But here we will discuss in detail the case where the pixels are mass tiles with uniform but adjustable surface density [112,139].

Consider a lens made up of N pixels each with mass profile $\kappa_n f_n(\vec{\theta})$. Here κ_n is an adjustable parameter.[5] Let $Q_n(\vec{\theta})$ be the integral of $\nabla^{-2} f_n(\vec{\theta})$ over the n-th pixel. In other words, let $\kappa_n Q_n(\vec{\theta})$ be the n-th pixel's contribution to the lens potential at $\vec{\theta}$. For square tiles or Gaussian tents Q_n is known but messy [112,2]. For harmonic components or other eigenfunctions of ∇^2, $Q_n(\vec{\theta})$ is simply proportional to $f_n(\vec{\theta})$. For a pixelated lens the arrival time surface (1.11) becomes

$$\tau(\vec{\theta}) = \tfrac{1}{2}\vec{\theta}^2 - \vec{\theta}\cdot\vec{\beta} - \sum_n \kappa_n Q_n(\vec{\theta}) \qquad (1.47)$$

where we have discarded a $\tfrac{1}{2}\vec{\beta}^2$ term from (1.11) since it is constant over the surface.

We may now implement three kinds of observational constraints.

1. Image positions: an image observed at $\vec{\theta_i}$ implies

$$\vec{\nabla}\tau(\vec{\theta_i}) = 0. \qquad (1.48)$$

---

[5] Hence we deprecate the alternative name 'non-parametric' for this method, favoring 'free-form' or 'pixelated'.

(We can safely neglect the uncertainty in $\vec{\theta}_i$, since image astrometry is typically at the mas level). A multiple-image system derives from the same $\vec{\beta}$, but that $\vec{\beta}$ is unknown. So each such system introduces $2(\langle\text{images}\rangle - \langle\text{sources}\rangle)$ constraints.

2. A time-delay measurement between images at $\vec{\theta}_i$ and $\vec{\theta}_j$ implies

$$\tau(\vec{\theta}_i) - \tau(\vec{\theta}_j) = h\,T_0^{-1} \times \langle\text{obs delay}\rangle. \tag{1.49}$$

In a quad there may be two or three independent time delays.

3. Tensor magnifications are measured from images of a multiple-component source. The implied constraints can be included simply by treating the images of separate components as independent image systems. A scalar magnification, or simple flux ratio, cannot be included in this way; however, flux ratios are sensitive to microlensing and thus less suitable for constraining macro-models.

All these constraints are linear in the unknowns $\kappa_n(\vec{\theta})$ and $\vec{\beta}$. Schematically, we may write

$$\begin{pmatrix} \text{Lensing} \\ \text{data} \end{pmatrix} =$$

$$\begin{pmatrix} \text{A} & \text{messy} & \text{but} & \text{linear} & \text{operator} \\ \text{also} & \text{involving} & \text{the} & \text{same} & \text{data} \end{pmatrix} \begin{pmatrix} \text{The} \\ \text{lens's} \\ \text{projected} \\ \text{mass} \\ \text{distribution} \end{pmatrix}$$

Note the un-square matrix: there are many more pixels than data, i.e., the reconstruction problem is highly underdetermined.

It is easy to find lens profiles formally consistent with observational constraints as above, but most of them will not look anything like galaxies. We now try to exclude the latter with additional constraints — in Bayesian terminology we apply a prior. A reasonable prior is the following.

- $\kappa_n \geq 0$,
- 180° rotation symmetry (optional),
- density gradient $\leq 45°$,
- $\kappa_n \leq 2\langle\text{average of neighbors}\rangle$, except for the central pixel,
- κ steeper than $|\vec{\theta}|^{-1/2}$, based on stellar dynamics evidence that the 3D density in galaxies is steeper than $r^{-1.5}$.

The observational and prior constraints confine allowed lenses to a convex polyhedron in the space of $(\kappa_1, \kappa_2, \ldots, \kappa_N)$. This can now be searched by a random-walk technique, yielding an ensemble of models. And then one can estimate h or whatever from that ensemble [139].

There are three caveats associated with this technique.

- The results depend on the prior, and the above prior certainly has too little information. But it has at least the advantage that uncertainty estimates will be conservative.
- Having the mass on tiles means that the models cannot have central density cusps, contrary to what galaxies are thought to have. But far from the center, this is not an issue because of the monopole degeneracy.
- There is too much pixel-scale structure. This is not an issue for h, but if one wanted input for microlensing computations then the pixel-scale structure needs to be smoothed out.

Four Well-Known Systems. We now describe some new results obtained by one of us (PS) with L.L.R. Williams, on four lenses. For each lens, there was an ensemble of 200 models. The κ_n controlled ~ 500 mass tiles, plus a parametrized external shear.

Q 0957+561 The reconstructions use the positions, tensor magnifications [39], and time delays [65] of the quasar, and another double formed by a knot in the quasar's host galaxy [13]. Figure 1.22 shows (i) the image configuration and schematic saddle-point contours for the quasar, (ii) the ensemble-average mass map, (iii) the h values from each model in the ensemble plotted against the radial-profile index of that model between the innermost and outermost images, and (iv) a histogram of the h values from the ensemble. The radial profile index corresponds roughly to $\alpha - 1$ for small values of α as defined in (1.37).

Two things are very noticeable in Fig. 1.22. The first is the largeness of the uncertainty in h; even in this lens with VLBI structure giving tensor magnifications and a time delay accurate to 1%, h values between 0.5 and 1 are all admissible. The second noticeable thing is the near-proportionality of h and the radial index, and it points us to the dominant source of the uncertainty: changing the radial index is almost equivalent to applying the mass disk degeneracy transformation, which rescales the time delays, and hence h, while having no effect on image positions or tensor magnifications.

PG 1115+080 Here the reconstructions use only image positions of the quasar (an inclined quad) and time delays [116,9]. Figure 1.23 shows the results, following the same plan as before. Again h has a large uncertainty, but is strongly correlated with the radial profile. But the distribution of h values is on average lower than for 0957+561. This promises improved results if results for several lenses are combined.

B 1608+656 The reconstructions from this inclined quad use image positions and time delays [34]. The lensing galaxy in this system appears to be a binary [18], so the 180°-rotation symmetry is not imposed. Figure 1.24 shows the results. It is interesting that the mass profile comes out elongated towards the visible second galaxy, even though the reconstructions had no information about the light from the lensing galaxies.

RX J0911+055 This is a short-axis quad with a preliminary time delay [47], and Fig. 1.25 shows the results.

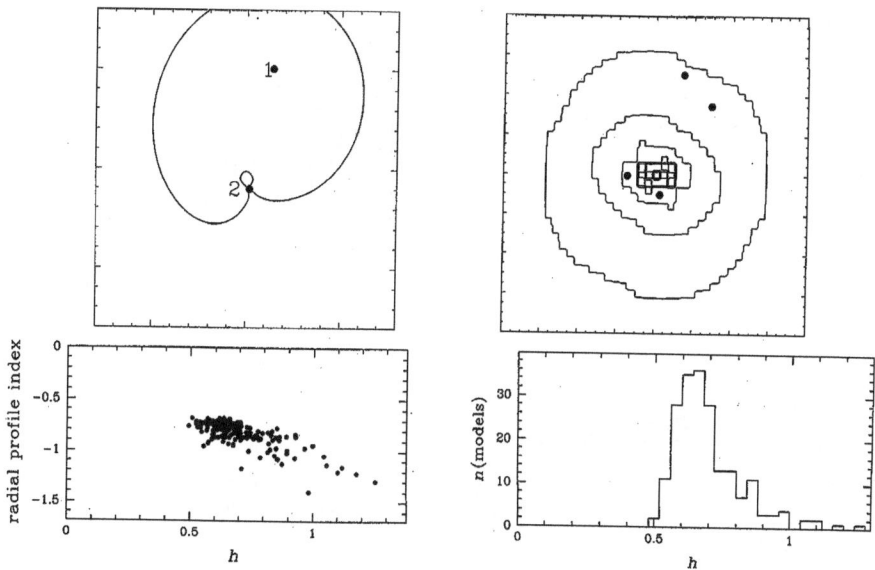

Fig. 1.22. Models of Q 0957+561. **Upper left:** schematic image configuration and saddle-point contours for the quasar. **Upper right:** ensemble-average reconstructed mass map; contours are $\kappa = \frac{1}{3}, \frac{2}{3}, \dots$ The four images marked are the quasar double and another double from a knot in the host galaxy. **Lower left:** h against radial index for all 200 models in the ensemble. **Lower right:** Histogram of h from the ensemble.

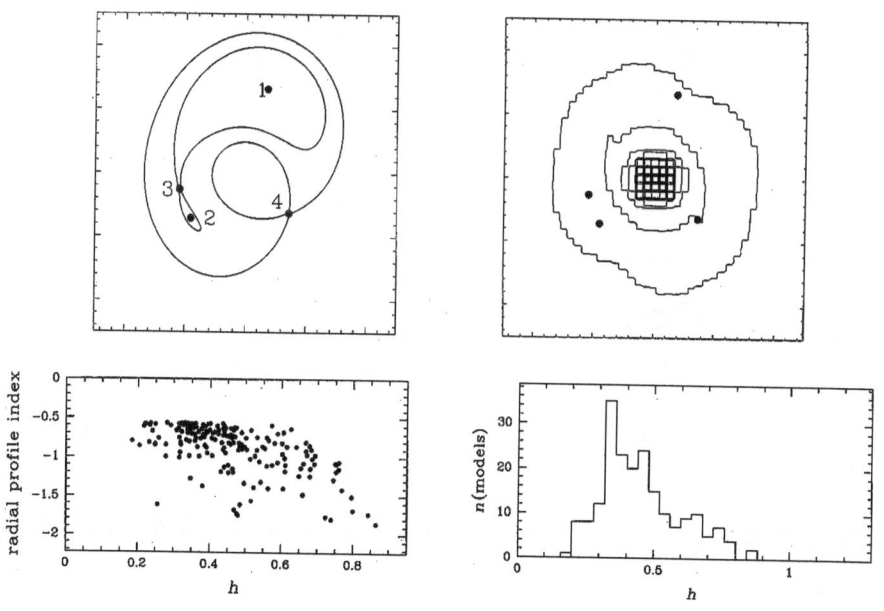

Fig. 1.23. Models of PG 1115+080. Panels arranged as in Fig. 1.22.

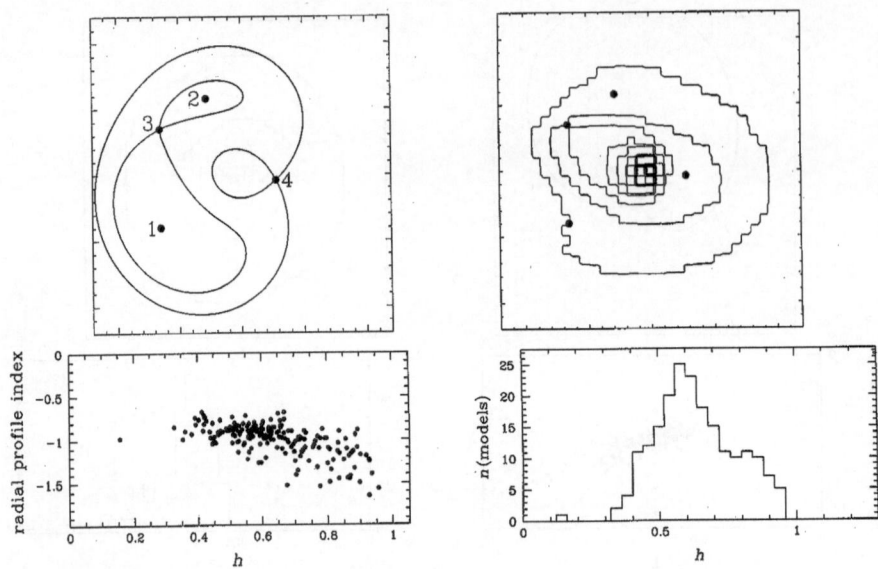

Fig. 1.24. Models of B 1608+656. Panels arranged as in Figs. 1.22 and 1.23.

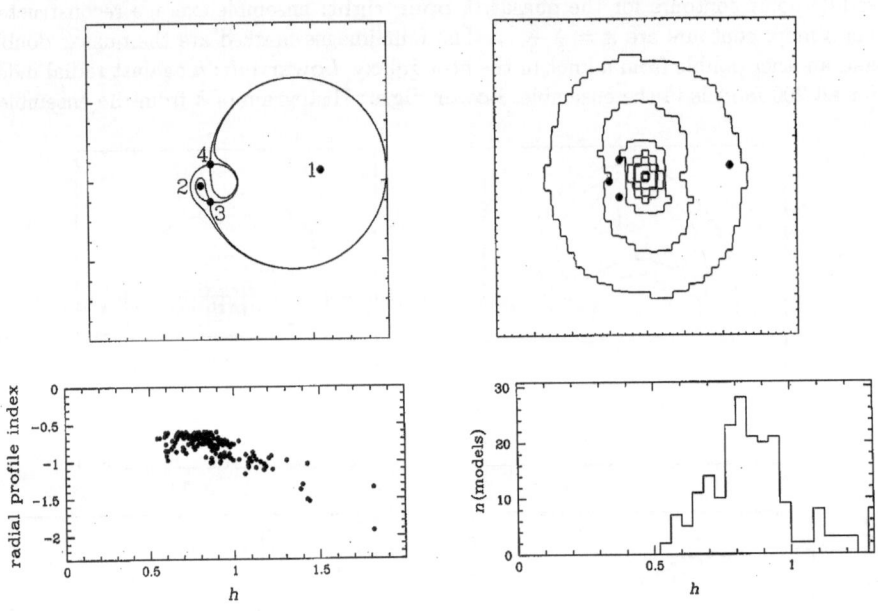

Fig. 1.25. Models of RX J0911+055. Panels arranged as in Figs. 1.22, 1.23, and 1.24.

Fig. 1.26. Ring and arc models resulting from plotting arrival-time surfaces with dense contours. **Left panel:** PG 1115+080 with contours 80 min apart; **middle panel:** B 1608+656 with contours 2 hr apart; **right panel:** 0957+561 with contours 1 day apart.

Ring and Arcs. The models described above are designed to fit images of (one or more) point sources. But having produced a model, one can check what sort of image it produces for extended sources. For a source with a conical or tent profile for brightness, the image is particularly easy to produce. We just have to make a dense contour map of the arrival-time surface for the center of the source and then view this map from a distance so that the contour lines blur into a grayscale [139,114]; the ratio

$$\frac{\tau\text{-spacing between contours}}{\text{thickness of contour lines}}$$

is proportional to the source size.

Figure 1.26 shows ring and arc images generated in this way from the ensemble-average models of PG 1115+080, B 1608+656, and Q 0957+561. These may be compared with published images of observed rings [50,18,57]. For PG 1115+080 and B 1608+656 the model and observed rings overlay extremely well. (Recall that the modeling procedure used no ring/arc data.) For Q 0957+561 the agreement is not so good: this may indicate simply that the models are less good, or it may indicate that the observed arc is the image not of the quasar host galaxy but another galaxy, possibly at different redshift.

Combined h Results. Returning to estimates of h, in Fig. 1.27 we show the result of combining the h distributions from all four systems above.

The combined result is

$$H_0 = 64^{+4}_{-4} \quad (68\%)$$
$$= 64^{+12}_{-6} \quad (90\%)$$

The reference cosmology is $(\Omega_\Lambda=0.7, \Omega_m=0.3)$; for other cosmologies the numbers would change by 5–10%.

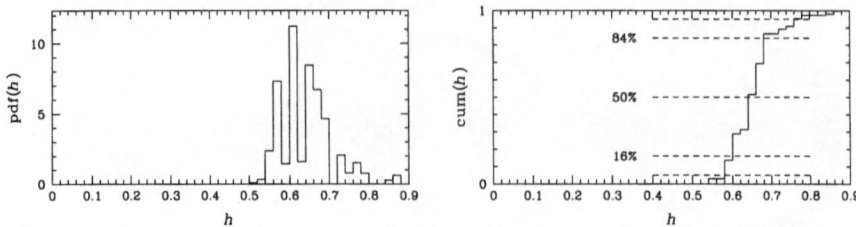

Fig. 1.27. Combined h results from Q 0957+561, PG 1115+080, B 1608+656, and RX J0911+055: histogram on the left and cumulative plot on the right.

1.4 Summary and Future Prospects

The present article concentrates on some selected aspects of quasar lensing, and in particular on their use for determining H_0. Lensed quasars have the advantage over other methods that they do not rely on the knowledge of any standard candle. The disadvantage is that precise modeling of the potential well responsible for the lensing effect is required.

It has been shown that, with present day instrumentation and efficient post-processing techniques, "mass production" of time-delays is possible, even using 2m class telescopes under average seeing conditions [21]. A typical precision on a time-delay determination is of the order of 10%, sometimes better. However, in many cases, most of the error on H_0 comes from the lens models used to convert the time-delay into H_0. The problem can be overcome in two ways: (1) by using any prior knowledge available on (lens) galaxies and, (2) by improving the observations to constrain better the gravitational potential (main lens and any intervening cluster/group) in each individual lensed quasar.

The effectiveness of quasar lensing in producing a competitive value for H_0 therefore depends on our knowledge of the physics of galaxies in general. Gravitational lensing itself should be able to set suitable constraints on galaxy mass profiles, for example through the statistical study of galaxy halos using galaxy-galaxy lensing (see Chap. 3). The development of two-dimensional spectrographs used to derive the full velocity field for many galaxies of all types will also yield important clues to the detailed mass distribution in galaxies. Both methods, direct or indirect, should constrain the degree of concentration of the mass in galaxies and the extent and shape of dark matter halos, two quantities which are often unconstrained in present days lens models and which imply the exploration of huge parameter spaces followed by a choice of a "best" model or a best family of models.

Improving the observations of individual lensed systems is also important. Lensing galaxies have an effect not only on quasars but also on galaxies in the vicinity of quasars. They should be seen under the form of arcs or arclets, as long as the angular resolution and depth are sufficient. The Advanced Camera for Surveys (ACS), on board of the HST, shall provide us at least with depth, hence with more background sources susceptible to be lensed, just as is the quasar. Since we usually observe only 2 or 4 quasar images, observing even a few arclets is a significant constraint for the lens model. In addition, with the depth of the ACS, most lensed quasars should show their distorted host galaxy, and bring even more constraints on the models. Constraining lens models using many arclets will probably become an efficient method with the launch of the Next Generation Space Telescope.

Measuring H_0 is not the only application of quasar lensing. Once adequate observational constrains are available, or even assuming H_0 can be measured independently by other methods, one shall use lensing to map the mass distribution in lensing galaxies and to infer basic parameters on the structure of quasars, using the chromatic variations due to microlensing events. Spectrophotometric monitoring is the next obvious observational step in the field, in order to enable such applications.

Whether we will learn about the mass distribution in galaxies once H_0 is measured by other means, or the opposite, will depends on the speed of the progress made in the fields of the physics of galaxies, galaxy-galaxy lensing and on the possible discoveries of new methods to infer H_0 with a high precision.

1.5 Inventory of Known Systems

The numerous quasar surveys carried out to date and others still under way have led to the discovery of many lensed systems. Consequently, it is becoming increasingly difficult to keep track of all new lenses discovered. We try here to provide the reader with a list of all known cases; we apologize in advance to those who will not see their favorite system, probably because it has been too recently discovered. Note also that we list only lensed quasars. There are other cases of multiply imaged distant galaxies, discovered for example in HST deep fields (see for example [10]). Basic information such as coordinates, source and lens redshifts are given, together with the reference of the discovery paper. When several references are listed, the first ones corresponds to the discovery paper, and the others to the time delay measurement, when available. Time delays are given relative to the leading image. For example, $\Delta t(BA)$ means that image B is the leading image. Note finally that most objects have been observed or will be observed with the HST, either in the context or individual observing programs or through the CASTLE Survey whose main results are summarized at http://cfa-www.harvard.edu/glensdata/.

Table 1.1. List of confirmed doubles.

Object	Coords (2000)	Redshifts	Notes
Q 0142-100	α: 01h 45m 16.50s	z_s=2.72	
Surdej et al. [126]	δ: −09d 45m 17.00s	z_l=0.49	
CTQ 414	α: 01h 58m 41.44s	z_s=1.29	
Morgan et al. [84]	δ: −43d 25m 04.20s	z_l=?	
B 0218+357	α: 02h 21m 05.48s	z_s=0.96	Δt(BA) = 10.5
O'dea et al. [31]	δ: +35d 56m 13.78s	z_l=0.68	± 0.4 days
Biggs et al. [14]			
HE 0512-3329	α: 05h 14m 10.78s	z_s=1.57	
Gregg et al. [41]	δ: −33d 26m 22.50s	z_l=0.93(?)	
CLASS B0739+366	α: 07h 42m 51.20s	z_s=?	
Marlow et al. [82]	δ: +36d 34m 43.70s	z_l=?	
MG 0751+2716	α: 07h 51m 41.46s	z_s=3.20	Ring
Lehar et al. [71]	δ: +27d 16m 31.35s	z_l=0.35	
HS 0818+1227	α: 08h 21m 39.10s	z_s=3.12	
Hagen & Reimers [43]	δ: +12d 17m 29.00s	z_l=0.39	
APM 08279+5255	α: 08h 31m 44.94s	z_s=3.87	
Irwin et al. [51]	δ: +52d 45m 17.70s	z_l=?	
SBS 0909+532	α: 09h 13m 01.05s	z_s=1.38	
Kochanek et al. [60]	δ: +52d 59m 28.83s	z_l=0.83	
RXJ 0921+4528	α: 09h 21m 12.81s	z_s=1.66	
	δ: +45d 29m 04.40s	z_l=0.31	
FBQ 0951+2635	α: 09h 51m 22.57s	z_s=1.24	
Schechter et al. [117]	δ: +26d 35m 14.10s	z_l=?	
BRI 0952-0115	α: 09h 55m 00.01s	z_s=4.5	
McMahon & Irwin [80]	δ: −01d 30m 05.00s	z_l=?	
Q 0957+561	α: 10h 01m 20.78s	z_s=1.41	Δt(BA) = 417
Walsh et al. [133]	δ: +55d 53m 49.40s	z_l=0.36	± 3 days
Kundić et al. [65]			
LBQS 1009-0252	α: 10h 12m 15.71s	z_s=2.74	
Surdej et al. [127]	δ: −03d 07m 02.00s	z_l=?	
Q 1017-207	α: 10h 17m 24.13s	z_s=2.55	
Claeskens et al. [27]	δ: −20d 47m 00.40s	z_l=?	
FSC 10214+4724	α: 10h 24m 37.58s	z_s=2.29	Ring
Graham & Liu [40]	δ: +47d 09m 07.20s	z_l=?	
B 1030+074	α: 10h 33m 34.08s	z_s=1.54	
Xanthopoulos et al. [152]	δ: +07d 11m 25.50s	z_l=0.60	
HE 1104-1805	α: 11h 06m 33.45s	z_s=2.32	Δt(AB) = 260
Wisotzki et al. [140]	δ: −18d 21m 24.20s	z_l=0.73	± 90 days
Wisotzki et al. [143]			
B 1127+385	α: 11h 30m 00.13s	z_s=?	
Koopmans et al. [63]	δ: +38d 12m 03.10s	z_l=?	

Table 1.2. List of confirmed doubles (continued)

Object	Coords (2000)	Redshifts	Notes
MG 1131+0456	α: 11h 31m 56.48s	z_s=?	
Hewitt et al. [45]	δ: +04d 55m 49.80s	z_l=0.84	
B 1152+199	α: 11h 55m 18.37s	z_s=1.02	
Myers et al. [89]	δ: +19d 39m 40.39s	z_l=0.44	
Q 1208+1011	α: 12h 10m 57.16s	z_s=3.80	
Magain et al. [78]	δ: +09d 54m 25.60s	z_l=?	
SBS 1520+530	α: 15h 21m 44.83s	z_s=1.86	Δt(BA) = 130
Chavushyan et al. [26]	δ: +52d 54m 48.60s	z_l=0.71	$\pm$ 6 days
Burud et al. [23]			
MG 1549+3047	α: 15h 49m 12.37s	z_s=?	Ring
Lehar et al. [69]	δ: +30d 47m 16.60s	z_l=0.11	
B 1600+434	α: 16h 01m 40.45s	z_s=1.59	Δt(BA) = 51
Jackson et al. [52]	δ: +43d 16m 47.80s	z_l=0.41	$\pm$ 4 days (radio)
Burud et al. [20]			Δt(BA) = 47
Koopmans et al. [64]			$\pm$ 11 days (radio)
PMN J1632-0033	α: 16h 32m 55.98s	z_s=3.42	
Winn et al. [146]	δ: −00d 33m 04.50s	z_l=?	
FBS 1633+3134	α: 16h 33m 48.99s	z_s=1.52	
Morgan et al. [85]	δ: +31d 34m 11.90s	z_l=?	
MG 1654+1346	α: 16h 54m 41.83s	z_s=1.74	Ring
Langston et al. [66]	δ: +13d 46m 22.00s	z_l=0.25	
PKS 1830-211	α: 18h 33m 39.94s	z_s=2.51	Ring
Subrahmanyan et al. [125]	δ: −21d 03m 39.70s	z_l=0.89	Δt(BA) = 26
Lovell et al. [76]			$\pm$ 8 days
PMN J1838-3427	α: 18h 38m 28.50s	z_s=2.78	
Winn et al. [144]	δ: −34d 27m 41.60s	z_l=?	
B 1938+666	α: 19h 38m 25.19s	z_s=?	Full ring
Rhoads et al. [107]	δ: +66d 48m 52.20s	z_l=0.88	
PMN J2004-1349	α: 20h 04m 07.02s	z_s=?	
Winn et al. [145]	δ: −13d 49m 31.65s	z_l=?	
B 2114+022	α: 21h 16m 50.75s	z_s=?	
	δ: +02d 25m 46.90s	z_l=0.32/0.59	
HE 2149-2745	α: 21h 52m 07.44s	z_s=2.03	Δt(BA) = 103
Wisotski et al. [141]	δ: −27d 31m 50.20s	z_l=0.49	$\pm$ 12 days
Burud et al. [22]			
B 2319+051	α: 23h 21m 40.80s	z_s=?	
Rusin et al. [111]	δ: +05d 27m 36.40s	z_l=0.62	

Table 1.3. List of central quads.

Object	Coords (2000)	Redshifts	Notes
CLASS B0128+437	α: 01h 31m 16.26s	z_s=?	
Phillips et al. [99]	δ: +43d 58m 18.00s	z_l=?	
HST 1411+5211	α: 14h 11m 19.60s	z_s=2.81	
Fischer et al. [36]	δ: +52d 11m 29.00s	z_l=0.46	
H 1413+117	α: 14h 15m 46.40s	z_s=2.55	
Magain et al. [77]	δ: +11d 29m 41.40s	z_l=?	
HST 14176+5226	α: 14h 17m 36.51s	z_s=3.4	
Ratnatunga et al. [102]	δ: +52d 26m 40.00s	z_l=0.81	
B 1555+375	α: 15h 57m 11.93s	z_s=?	
Marlow al. [81]	δ: +37d 21m 35.90s	z_l=?	
Q 2237+0305	α: 22h 40m 30.34s	z_s=1.69	
Huchra et al. [49]	δ: +03d 21m 28.80s	z_l=0.04	

Table 1.4. List of short axis quads.

Object	Coords (2000)	Redshifts	Notes
B 1422+231	α: 14h 24m 38.09s	z_s=3.62	
Patnaik et al. [95]	δ: +22d 56m 00.60s	z_l=0.34	

Table 1.5. List of long axis quads.

Object	Coords (2000)	Redshifts	Notes
RXJ 0911.4+0551	α: 09h 11m 27.50s	z_s=2.8	Δt(BA) = 146
Bade et al. [7]	δ: +05d 50m 52.00s	z_l=0.77?	$\pm$ 8 days
Hjorth et al. [48]			
HST 12531-2914	α: 12h 53m 06.70s	z_s=?	
Ratnatunga et al. [102]	δ: $-$29d 14m 30.00s	z_l=?	
B 2045+265	α: 20h 47m 20.35s	z_s=1.28	
Fassnacht et al. [35]	δ: +26d 44m 01.20s	z_l=0.87	

Table 1.6. List of inclined quads.

Object	Coords (2000)	Redshifts	Notes
0047-2808	α: 00h 49m 41.89s	z_s=3.60	
Warren et al. [137]	δ: $-$27d 52m 25.70s	z_l=0.49	
HE 0230-2130	α: 02h 32m 33.10s	z_s=2.16	
Wisotzki et al. [143]	δ: $-$21d 17m 26.00s	z_l=?	
MG 0414+0534	α: 04h 14m 37.73s	z_s=2.64	
Hewitt et al. [46]	δ: +05d 34m 44.30s	z_l=0.96	
B 0712+472	α: 07h 16m 03.58s	z_s=1.34	
Jackson et al. [53]	δ: +47d 08m 50.00s	z_l=0.41	
PG 1115+080	α: 11h 18m 17.00s	z_s=1.72	$\Delta t(AB) = 11.7$
Weymann et al. [138]	δ: +07d 45m 57.70s	z_l=0.31	$\pm$ 1.2 days
Schechter et al [117]			$\Delta t(CB) = 25.0$
			$\pm$ 1.6 days
B 1608+656	α: 16h 09m 13.96s	z_s=1.39	$\Delta t(BA) = 31$
Myers et al. [88]	δ: +65d 32m 29.00s	z_l=0.63	$\pm$ 7 days
Fassnacht et al. [34]			$\Delta t(BC) = 36$
			$\pm$ 7 days
			$\Delta t(BD) = 76$
			$\pm$ 10 days
MG 2016+112	α: 20h 19m 18.15s	z_s=3.27	
Lawrence et al. [68]	δ: +11d 27m 08.30s	z_l=1.01	

Table 1.7. List of systems with more than four images.

Object	Coords (2000)	Redshifts	Notes
B 1359+154	α: 14h 01m 35.55s	z_s=3.24	6 images
Myers et al [89]	δ: +15d 13m 25.60s	z_l=?	
B 1933+507	α: 19h 34m 30.95s	z_s=2.63	10 images
Sykes et al. [129]	δ: +50d 25m 23.60s	z_l=0.76	

References

1. H.M. AbdelSalam: D.Phil. thesis, Oxford (1998)
2. H.M. AbdelSalam, P. Saha, L.L.R. Williams: NewAR **42**, 157 (1998)
3. E. Agol, J. Krolik: ApJ **524**, 49 (1999)
4. C. Alard, R. Lupton: ApJ **503**, 325 (1998)
5. M.-C. Angonin-Willaime, G. Soucail, C. Vanderriest: A&A **291**, 411 (1994)
6. M.-C. Angonin-Willaime, C. Vanderriest, F. Courbin: A&A **347**, 434 (1999)
7. N. Bade, J. Siebert, S. Lopez, et al.: A&A **317**, L13 (1997)
8. N.T.J Bailey: *The Mathematical Approach to Biology and Medicine*, New York, Wiley (1967)

9. R. Barkana: ApJ **489**, 21 (1997)
10. R. Barkana, R.D. Blandford, D.W. Hogg: ApJ **513**, L91 (1999)
11. R. Bender, C. Moellenhoff: A&A **177**, 71 (1987)
12. G. Bernstein, P. Fischer, T.J. Tyson, et al.: ApJ **483**, L79 (1997)
13. G. Bernstein, P. Fischer: AJ **118**, 14 (1999)
14. A.D. Biggs, I.W.A. Browne, P. Helbig, et al.: MNRAS **304**, 349 (1999)
15. J. Binney, S. Tremaine: *Galactic Dynamics*, Princeton University Press (1987)
16. R.D. Blandford, R. Narayan: ApJ **310**, 568 (1986)
17. R.D. Blandford, H. Netzer, L. Woltjer, et al.: *Active galactic nuclei*, Springer-Verlag (1990)S
18. R.D. Blandford, G. Surpi: *Gravitational Lensing: Recent Progress and Future Goals*, ASP Conf. Series, Vol 237 (eds. Brainerd & Kochanek), p103 (2001)
19. I. Burud, F. Courbin, C. Lidman, et al.: ApJ **501**, L5 (1998)
20. I. Burud, J. Hjorth, A.O. Jaunsen, et al.: ApJ **544**, 117 (2000)
21. I. Burud: D. Phil. thesis, Université de Liège, Belgium (2001)
22. I. Burud, F. Courbin, P. Magain, et al.: A&A **383**, 71 (2002)
23. I. Burud, J. Hjorth, F. Courbin, et al.: A&A, in press (2002)
24. R.M. Campbell, J. Lehar, B.E. Corey, et al.: AJ **110**, 2566 (1995)
25. K. Chang, S. Refsdal: Nature **282**, 561 (1979)
26. V.H. Chavushyan, V.V. Vlasyuk, J.A. Stepanian, et al.: A&A **318**, L67 (1997)
27. J.-F. Claeskens, J. Surdej, M. Remy: A&A **305**, L9 (1996)
28. F. Courbin, P. Magain, C.R. Keeton, et al.: A&A **324**, L1 (1997)
29. F. Courbin, P. Magain, M. Kirkove, et al.: ApJ **529**, 1136 (2000)
30. F. Courbin, C. Lidman, G. Meylan, et al.: A&A **360**, 853 (2000)
31. C.P. O'Dea, S.A. Baum, C. Stanghellini, et al.: AJ **104**, 1320 (1992)
32. E.E. Falco, M.V. Gorenstein, I.I. Shapiro: ApJ 289, L1 (1985)
33. C.D. Fassnacht, J.G. Cohen: AJ **115**, 377 (1998)
34. C.D. Fassnacht, T.J. Pearson, A. Readhead, et al.: ApJ **527**, 498 (1999)
35. C.D. Fassnacht, R.D. Blandford, J.G. Cohen, et al.: AJ **117**, 658 (1999)
36. P. Fischer, D. Schade, F. Barrientos: ApJ **503**, L127 (1998)
37. M. Fukugita, T. Futamase, M. Kasai, et al.: ApJ **393**, 3 (1992)
38. R. Florentin-Nielsen, K. Augustesen: IAUC **3945**, 2 (1984)
39. M.A. Garrett, R.J. Calder, R.W. Porcas, et al.: MNRAS **270**, 457 (1994)
40. J.R. Graham, M.C. Liu: ApJ **449**, L29 (1995)
41. M. D. Gregg, L. Wisotzki, R.H. Becker, et al.: AJ **119**, 2535 (2000)
42. D.B. Haarsma, J.N. Hewitt, J. Lehar, et al.: ApJ **510**, 64 (1999)
43. H.-J. Hagen, D. Reimers: A&A **357**, L29 (2000)
44. E.K. Hege, E.N. Hubbard, P.A. Strittmatter, et al.: ApJ **248**, L1 (1981)
45. J.N. Hewitt, E.L. Turner, D.P. Schneider, et al.: Nature **333**, 573 (1988)
46. J.N. Hewitt, E.L. Turner, C.R. Lawrence, et al.: AJ **104**, 968 (1992)
47. J. Hjorth, I. Burud, A.O. Jaunsen, et al.: *Gravitational Lensing: Recent Progress and Future Goals*, ASP Conf. Series, Vol 237 (eds. Brainerd & Kochanek), p125 (2001)
48. J. Hjorth, I. Burud, A.O. Jaunsen, et al.: ApJ **572**, L11 (2002)
49. J. Huchra, M. Gorenstein, S. Kent, et al.: AJ **90**, 691 (1985)
50. C.D. Impey, E.E. Falco, C.S. Kochanek, et al.: ApJ **509**, 551 (1998)
51. M.J. Irwin, R.A. Ibata, G.F. Lewis, et al.: ApJ **505**, 529 (1998)
52. N. Jackson, A.G. De Bruyn, S. Myers, et al.: MNRAS **274**, L25 (1995)
53. N. Jackson, S. Nair, I.W.A. Browne, et al.: MNRAS **296**, 483 (1998)
54. A.O. Jaunsen, J. Hjorth: A&A **317**, L39 (1997)

55. R. Kayser, S. Refsdal: A&A **128**, 156 (1983)
56. C. R. Keeton, C. S. Kochanek: ApJ **487**, 42 (1997)
57. C.R. Keeton, E.E. Falco, C.D. Impey, et al.: ApJ **542**, 74 (2000)
58. J.-P. Kneib, J.G. Cohen, J. Hjorth: ApJ **544**, 35 (2000)
59. C.S. Kochanek, R. Narayan: ApJ **401**, 461 (1992)
60. C.S. Kochanek, E.E. Falco, R. Schild, et al.: ApJ **479**, 678 (1997)
61. C.S. Kochanek, E.E. Falco, C.D. Impey, et al.: *The CASTLE Survey*, http://cfa-www.harvard.edu/glensdata/
62. C.S. Kochanek, C.R. Keeton, B.A. McLeod: ApJ, **547**, 50 (2001)
63. L. V. E. Koopmans, A.G. Bruyn, D.R. De Marlow, et al.: MNRAS **303**, 727 (1999)
64. L.V.E. Koopmans, A.G. de Bruyn, E. Xanthopoulos, et al.: A&A, **356**, 391 (2000)
65. T. Kundic, E.L. Turner, W.N. Colley, et al.: ApJ **482**, 75 (1997)
66. G.I. Langston, D.P. Schneider, S. Conner, et al.: AJ **97**, 1283 (1989)
67. C.R. Lawrence, D.P. Schneider, D.P. Schmidt, et al.: Science **223**, 46 (1984)
68. C. R. Lawrence, D.P. Schneider, M. Schmidet, et al.: Science **223**, 46 (1984)
69. J. Lehar, G.I. Langston, A.D. Silber, et al.: AJ **105**, 847 (1993)
70. J. Lehar, A.J. Cooke, C.R. Lawrence, et al.: AJ **111**, 1812 (1996)
71. J. Lehar, B.F. Burke, S.R. Conner, et al.: AJ **114**, 58 (1997)
72. J. Lehar, E.E. Falco, C.S. Kochanek, et al.: ApJ **536**, 584 (2000)
73. G.F. Lewis, M.J. Irwin: MNRAS **276**, 103 (1996)
74. C. Lidman, F. Courbin, G. Meylan, et al.: ApJ **514**, L57 (1999)
75. C. Lidman, F. Courbin, J.-P. Kneib, et al.: A&A **364**, L62 (2000)
76. J.E.J. Lovell, D.L. Jauncey, J.E. Reynolds, et al.: ApJ **508**, L51 (1998)
77. P. Magain, J. Surdej, J.-P. Swings, et al.: Nature **334**, 327 (1988)
78. P. Magain, J. Surdej, C. Vanderriest: A&A **253**, L13 (1992)
79. P. Magain, F. Courbin, S. Sohy: ApJ **494**, 472 (1998)
80. R. McMahon, M. Irwin: Gemini **36**, 1 (1992)
81. D. R. Marlow, S.T. Myers, D. Rusin, et al.: AJ **118**, 654 (1999)
82. : D. R. Marlow, D. Rusin, M.A.J. Norbury, et al.: AJ **121**, 619 (2001)
83. B. Moore, N. Katz, G. Lake: IAUS **171**, 203 (1996)
84. N.D. Morgan, A. Dressler, J. Maza, et al.: AJ **118**, 1444 (1999)
85. N.D. Morgan, R.H. Becker, M.D. Gregg, et al.: AJ **121**, 611 (2001)
86. N.D. Morgan, G. Chartas, M. Malm, et al.: ApJ **555**, 1 (2001)
87. J. A. Munoz, E.E. Falco, C.S. Kochanek, et al.: ApJ **546**, 769 (2001)
88. S.T. Myers, C.D. Fassnacht, S.G. Djorgovski, et al.: ApJ **447**, L5 (1995)
89. S.T. Myers, D. Rusin, C.D. Fassnacht, et al.: AJ **117**, 2565 (1999)
90. I. Newton: *Optics 2nd Ed., query 1* (1704)
91. R. Nityananda: *Current Science* **59**, 1044 (1990)
92. A. Oscoz, E. Mediavilla, L.J. Goicoechea, et al.: ApJ **479**, L89 (1997)
93. R. Ostensen, S. Refsdal, R. Stabell, et al.: A&A **309**, 59 (1996)
94. B. Paczyński: ApJ **301**, 503 (1986)
95. A.R. Patnaik, I.W.A. Browne, D. Walsh, et al.: MNRAS **259**, 1 (1992)
96. J. Pelt, R. Schild, S. Refsdal: A&A **336**, 829 (1998)
97. J. Pelt, R. Kayser, S. Refsdal, T. Schramm: A&A **305**, 97 (1996)
98. J. Pelt, W. Hoff, R. Kayser, et al.: A&A **286**, 775 (1994)
99. P.M. Phillips, M.A. Norbury, L. V. E. Koopmans, et al.: MNRAS **319**, L7 (2000)
100. F.P. Pijpers: MNRAS **289**, 933 (1997)
101. W.H. Press, G.B. Rybicki, J.N.W. Hewitt: ApJ **385**, 416 (1992)
102. K.U. Ratnatunga, E.J. Ostrander, R.E Griffiths, et al.: ApJ **453**, L5 (1995)
103. S. Refsdal, R. Stabell, J. Pelt, et al.: A&A **360**, 10 (2000)

104. S. Refsdal: MNRAS **128**, 295 (1964)
105. S. Refsdal: MNRAS **128**, 307 (1964)
106. S. Refsdal: MNRAS **132**, 101 (1966)
107. J. Rhoads, S. Malhotra, T. Kundic: AJ **111**, 642 (1996)
108. D.H. Roberts, J. Lehar, J.N.W. Hewitt, et al.: Nature **352**, 43 (1991)
109. A.J. Romanowsky, C.S. Kochanek: ApJ **516**, 18 (1999)
110. E. Ros, J.C. Guirado, J.M. Marcaide, et al.: A&A **362**, 845 (2000)
111. D. Rusin, D.R. Marlow, M. Norbury, et al.: AJ **122**, 591 (2001)
112. P. Saha, L.L.R. Williams: MNRAS **292**, 148 (1997)
113. P. Saha: AJ **120**, 1654 (2000)
114. P. Saha, L.L.R. Williams: AJ **122**, 585 (2001)
115. N. Sanitt: Nature **234**, 199 (1971)
116. P.L. Schechter, C.D. Bailyn, B. Robert et al.: ApJ **475**, L85 (1997)
117. P. L. Schechter, M.D. Gregg, R.H. Becker, et al.: AJ **115**, 1371 (1998)
118. P. L. Schechter, J. Wambsganss: astro-ph/0204425 (2002)
119. R.E. Schild, B. Cholfin: ApJ **300**, 209 (1986)
120. R.E. Schild: AJ **100**, 1771 (1990)
121. M. Schmidt: Nature **197**, 1040 (1963)
122. P. Schneider: ApJ **319**, 9 (1987)
123. P. Schneider, C. Seitz: A&A **294**, 411 (1985)
124. I.I. Shapiro: Phys. Rev. Lett. **13**, 789 (1964)
125. R. Subrahmanyan, D. Narashima, A. Pramesh-Rao, et al.: MNRAS **246**, 263 (1990)
126. J. Surdej, P. Magain, J.-P. Swings, et al.: Nature **329**, 695 (1987)
127. J. Surdej, M. Remy, A. Smette et al.: *Proc. 31st Liege Int. Astroph. Coll. 'Gravitational Lenses in the Universe'*, p. 153 (1993)
128. A. Surdej: *Lensing bibliographie*: http://vela.astro.ulg.ac.be/themes/-extragal/gravlens/bibdat/engl/glb_homepage.html
129. C.M. Sykes, I.W.A. Browne, N.J. Jackson, et al.: MNRAS **301**, 310 (1998)
130. J.L. Tonry, C.S. Kochanek: AJ **117**, 2034 (1999)
131. C.S. Trotter, J.N. Winn, J.N. Hewitt: ApJ **535**, 671 (2000)
132. C. Vanderriest, J. Schneider, G. Herpe, et al.: A&A **215**, 1 (1989)
133. D. Walsh, R.F. Carwell, R.J Weymann: Nature **279**, 381 (1979)
134. J. Wambsganss, P. Schneider, B. Paczynski: ApJ **358**, L33 (1990)
135. J. Wambsganss: ApJ **386**, 19 (1992)
136. J. Wambsganss, R. Schmidt: NewAR **42**, 101 (2000)
137. S.J. Warren, P.C. Hewett, G.F. Lewis, et al.: MNRAS **278**, 139 (1996)
138. R. J. Weymann, D. Latham, J. Roger: Nature **285** 641 (1980)
139. L.L.R. Williams, P. Saha: AJ **119**, 439 (2000)
140. L. Wisotzki, T. Koehler, R. Kayser, et al.: A&A **278**, L15 (1993)
141. L. Wisotzki, T. Koehler, S. Lopez, et al.: A&A **315**, L405 (1996)
142. L. Wisotzki, O. Wucknitz, S. Lopez, et al.: A&A **339**, L73 (1998)
143. L. Wisotzki, N. Christlieb, M.C. Liu, et al.: A&A **348**, L41 (1999)
144. J. N. Winn, J.N. Hewitt, P.L. Schechter, et al.: AJ **120**, 286 (2000)
145. J.N. Winn, J.N. Hewitt, A.R. Patnaik, et al.: AJ **121**, 1223 (2001)
146. J.N. Winn, N.D. Morgan, J.N. Hewitt, et al.: AJ **123**, 10 (2002)
147. H.J. Witt: ApJ **403**, 530 (1993)
148. H.J. Witt, S. Mao, P.L. Schechter: ApJ **443**, 18 (1995)
149. P. R. Wozniak, A. Udalski, M. Szymanski, et al.: ApJ **540**, L65 (2000)
150. J.S.B. Wyithe, R.L. Webster, E.L. Turner, et al.: MNRAS **315**, 62 (2000)
151. J.S.B. Wyithe, R.L. Webster, E.L. Turner: MNRAS **318**, 1120 (2000)
152. E. Xanthopoulos, I.W.A. Browne, L.J. King: MNRAS **300**, 649 (1998)

2 Weak Lensing

David Wittman

Bell Laboratories, Lucent Technologies, Room 1E-414,
700 Mountain Avenue, Murray Hill, NJ 07974, USA

Abstract. In the preceding chapter, the effects of lensing were so strong as to leave an unmistakable imprint on a specific source, allowing a detailed treatment. However, only the densest regions of the universe are able to provide such a spectacular lensing effect. To study more representative regions of the universe, we must examine large numbers of sources statistically. This is the domain of weak lensing.

2.1 Introduction

2.1.1 Motivation

Weak lensing enables the direct study of mass in the universe. Lensing, weak or strong, provides a more direct probe of mass than other methods which rely on astrophysical assumptions (*e.g.* hydrostatic equilibrium in a galaxy cluster) or proxies (*e.g.* the galaxy distribution), and can potentially access a more redshift-independent sample of structures than can methods which depend on emitted light with its r^{-2} falloff. But strong lensing can be applied only to the centers of very dense mass concentrations. Weak lensing, in contrast, can be applied to the vast majority of the universe. It provides a direct probe of most areas of already-known mass concentrations, and a way to discover and study new mass concentrations which could potentially be dark. With sources covering a broad redshift range, it also has the potential to probe structure along the line of sight.

Specifically, we might expect weak lensing to answer these questions:

- Where are the overdensities in the universe ?
- Are they associated with clusters and groups of galaxies ? Does light trace mass in these systems ?
- How much do these systems contribute to Ω_m, the mean density of matter in the universe ?
- What is their mass function and how does that function evolve with redshift ? What does that imply for the dark energy equation of state ?
- What are the structures on larger scales (walls, voids, filaments) ?
- Is this structure comparable to that seen in cosmological simulations ? Which cosmology matches best ?
- What is the nature of dark matter ?
- Can observations of lensing put any constraints on alternative theories of gravity ?

Until recently, deep imaging on the scale required to answer the above questions with weak lensing was simply impractical. The development of large mosaics of CCDs has expanded the field greatly. The large data volume leads to ever-decreasing statistical errors, which means that very close attention must be paid to systematic errors and calibration issues. Weak lensing results must be carefully scrutinized and compared with those of other approaches with this in mind.

We start with a review of the basic concepts, the limits of weak lensing, and observational hurdles, and then address the above astrophysical questions.

2.1.2 Basics

The transition from strong to weak lensing can be seen at a glance in the simulation shown in Fig. 2.1. Over most of the field, no one galaxy is obviously lensed, yet the galaxies have a slight tendency to be oriented tangentially to the lens. We seek to exploit this effect to derive information about the lens, and perhaps about the weakly lensed sources as well.

Fig. 2.1. Simulated effects of a lens: source plane (left) and image plane (right). Most regions of the lens can be probed only with weak lensing. Real sources are not in a plane, but this does not dramatically affect the appearance. Real lenses, such as galaxy clusters, would obscure much of the strong-lensing region (see Color Plate).

We start with the *inverse magnification matrix* (see also Chap. 1)

$$M^{-1} = (1 - \kappa) \begin{pmatrix} 1 & 0 \\ 0 & 1 \end{pmatrix} + \gamma \begin{pmatrix} \cos 2\phi & \sin 2\phi \\ \sin 2\phi & -\cos 2\phi \end{pmatrix}, \qquad (2.1)$$

so called because it describes the change in source coordinates for an infinitesmal change in image coordinates, the inverse of the transformation undergone by the sources. This is (1.16) of Chap. 1, which derives M^{-1} and defines the quantities within. We repeat here that the *convergence* κ represents an isotropic magnification, and the *shear* γ represents a stretching in the direction ϕ. They are both related to physical properties of the lens as linear combinations of derivatives of the deflection angle. However, κ can be interpreted very simply as the projected mass density Σ divided by the critical density Σ_{crit}, while γ has no such straightforward interpretation. In fact, γ is *nonlocal*: its value at a given position on the sky depends on the mass distribution everywhere, not simply at that position. We will see this fact rear its ugly head in several places throughout this chapter. Shear is often written as a vector $\gamma_i = (\gamma \cos 2\phi, \gamma \sin 2\phi)$ or more succinctly as a complex quantity $\gamma e^{i2\phi}$.

Without multiple images of a source (as in the strong lensing case), we must have some independent knowledge of the sources if we are to measure magnification or shear. For example, if one source were a standard candle or ruler, the apparent magnitude or size of its image would immediately yield the magnification at that point. Of course, standard candles or rulers occur only in very special cases [17], so in practice we must analyze source *distributions*. We no longer get much information from a single source, and thus lose resolution; this is the tradeoff we must make for probing regions with weak tidal fields.

One source distribution that could be used in this way is $n(m)$, the number of galaxies as a function of apparent magnitude. In practice, this is difficult, because the measured slope of this distribution does not differ greatly from the critical slope at which equal numbers of galaxies are magnified into and out of a given magnitude bin, with no detectable change ($n \propto m^{0.4}$). There is enough difference to make some headway, but we would prefer to measure departures from zero rather than small changes in a large quantity.

The distribution of galaxy *shapes*, properly defined, does allow us to measure departures from zero. Approximate each source as an ellipse with position angle ϕ and (scalar) ellipticity $\epsilon = \frac{a^2 - b^2}{a^2 + b^2}$, where a and b are the semimajor and semiminor axes. Define a *vector ellipticity* $e_i = (\epsilon \cos 2\phi, \epsilon \sin 2\phi)$, or equivalently a *complex ellipticity* $\epsilon e^{i2\phi}$ (also called *polarization*). This encodes the position angle and scalar ellipticity into two quantities which are comparable to each other; the dependence on 2ϕ indicates invariance under rotation by $180°$. Fig. 2.2 gives a visual impression of ellipses in this space.

We can now quantify the visual impression of Fig. 2.1. In the absence of lensing, as in the left panel, galaxies are randomly oriented: The observed distribution of e_i is roughly Gaussian with zero mean and an rms of $\sigma_e \sim 0.3$. In the presence of lensing, as in the right panel, this distribution is no longer centered on zero, as long as we consider an appropriately-sized patch of sky. In fact, we will *assume* that any departure from zero mean must be due to lensing. We will examine the limits of this assumption in some detail later, but for now let us accept that on large enough scales, the cosmological principle demands it, and as a practical matter, we average over sources at a wide range of redshifts, which are too far apart physically to influence each other's alignment.

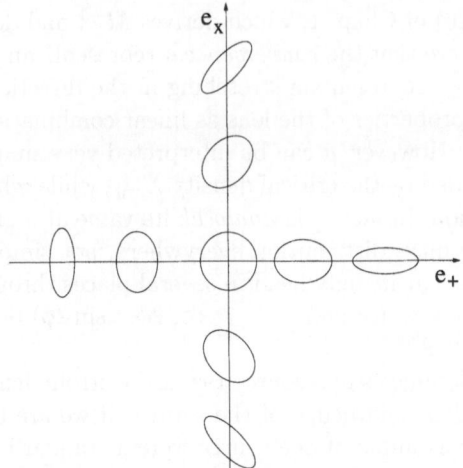

Fig. 2.2. A sequence of ellipses with various amounts of each ellipticity component. Inspired by the appearance of these ellipses, the two components are often labeled e_+ and $e_\times$.

The effect of the magnification matrix on the complex ellipticity can be computed if M is constant over a source. This is obviously not valid for very large sources or those near caustics, but it is valid for the vast majority of the sky and for typical sources with sizes of a few arcseconds. The result is that $\epsilon^I = \epsilon^S + \frac{\gamma}{1-\kappa}$, where superscripts indicate image and source planes [18]. We don't know any of these quantities for a single source, but we do know (or assume for now) that $\langle \epsilon^S \rangle = 0$, where brackets indicate averaging over many sources. Hence

$$\langle \epsilon^I \rangle = \langle \frac{\gamma}{1-\kappa} \rangle. \tag{2.2}$$

The quantity on the right is called the *reduced shear g*. A second approximation we can often make is that $\kappa \ll 1$, so that $\langle \epsilon^I \rangle = \langle \gamma \rangle$. This is called the *weak lensing limit*.

The fundamental limit to the accuracy with which we can measure γ in the weak lensing limit is *shape noise*, or the width of the source ellipticity distribution $\sigma_e \sim 0.3$. Averaging over n sources should decrease the uncertainty to $\frac{\sigma_e}{\sqrt{n}}$, but n is limited by the depth of the observations and the area over which we are willing to average γ; these tradeoffs are discussed below. Also note that knowledge of the shear alone is not strictly enough to infer mass distributions because of the *mass sheet degeneracy* [44,111], introduced in a different context in Chap. 1. This degeneracy arises because a uniform sheet of mass induces only magnification, not shear. Because the equations are linear, we could therefore add or subtract a mass sheet without affecting the shear. In practice, we can still answer many questions with shear alone, as discussed below.

2.1.3 Cosmology Dependence

Both convergence and shear scale as the combination of angular diameter distances $\frac{D_{LS}D_L}{D_S}$, or as the *distance ratio* $\frac{D_{LS}}{D_S}$ for a given lens. (Recall from the Chap. 1 that D_{LS}, D_L, and D_S are the angular diameter distances from lens to source, observer to lens, and observer to source, respectively. Note that $D_S \neq D_L + D_{LS}$; see [59] for a quick review and [101] for a thorough treatment of distance measures in cosmology). This cosmology-dependent quantity is plotted as a function of source redshift in Fig. 2.3 for several lens redshifts and two different cosmologies. In principle, this could be used to constrain the cosmology if source redshifts are known, and if the lens mass is known independently (the effects of a larger lens mass and a larger universe are degenerate). But this remains an unused cosmological test because lens parameters and source redshifts are usually poorly known. Usually, a cosmology is assumed and lens parameters are estimated using any available knowledge of source redshifts. Less often, a well-characterized lens is used to explore the source redshift distribution. However, source redshift distributions are usually quite broad, and weak lensing can only be used to estimate the *mean* distance ratio to a group of sources, which is not same as the distance ratio corresponding to the mean redshift. Section 2.1.4 deals with ways of estimating the mean distance ratio or otherwise accounting for a broad source redshift distribution.

Another way of viewing the same information is to fix the source redshift and plot this ratio as a function of lens redshift (Fig. 2.4). This reveals the relative importance of different structures along the line of sight and is often called the *lensing kernel* or *lensing efficiency*.

2.1.4 Applicability of Weak Lensing

As with all astrophysical tools, we must be aware of the limitations of weak lensing before plunging into results. They include the weak lensing approximation itself; mass sheet degeneracy if only shear is used; poor angular resolution because of its statistical nature; source redshift difficulties; and possible departures from the assumption of randomly oriented sources. We now examine these limits and ways of dealing with them.

Weak Lensing Approximation. The approximations that M is constant over each source and that $\kappa \ll 1$ cannot be applied when dealing with the centers of massive clusters and galaxies. Of course, analysis of such regions is not lacking— it is the topic of most of this book. Here we merely wish to mention work that has been done on combining weak and strong lensing information [1]. We also note that, where only the second approximation fails, (2.2) can be solved iteratively for κ.

Mass Sheet Degeneracy. Mass sheet degeneracy was a serious concern when fields of view were small and lens mass distributions extended well beyond the edges. Modern imagers now deliver fields of view $\sim 0.5°$ on a side (> 3 Mpc

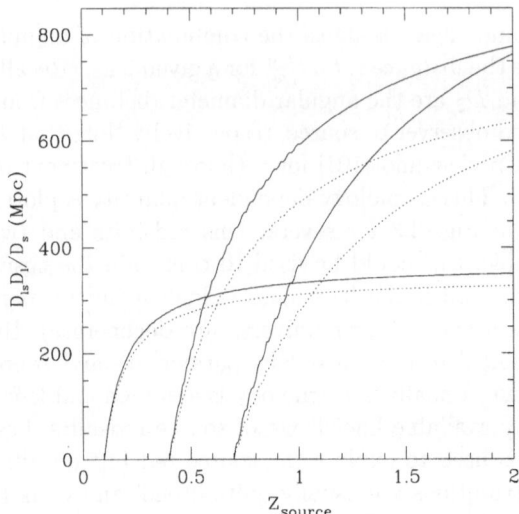

Fig. 2.3. $\frac{D_{\mathrm{LS}}D_{\mathrm{L}}}{D_{\mathrm{S}}}$ as a function of source redshift, for several lens redshifts (indicated by the intersections of the curves with the horizontal axis) and several cosmologies. The cosmologies are Λ-dominated (solid lines, $H_0 = 70$ km s^{-1} Mpc^{-1}, $\Omega_m = 0.3$, $\Omega_\Lambda = 0.7$) and open (dashed lines, $H_0 = 70$ km s^{-1} Mpc^{-1}, $\Omega_m = 0.4$, $\Omega_\Lambda = 0$). Each solid line is higher than its dashed counterpart, reflecting the larger size of the Λ-dominated universe. Although this quantity appears to be a sensitive test of the cosmology, it is degenerate with the lens mass.

radius for any lens at $z > 0.15$), so this concern has diminished. The degeneracy may also be broken by adding magnification information, which may come from strong lensing, or from a method called the depletion curve.

Magnification imposes two effects of opposite sign on the areal density of sources. Galaxies fainter than the detection limit (or any chosen brightness threshold) are amplified above the threshold, increasing the density of sources, but at the same time the angular separation between galaxies is stretched, decreasing the density of sources. The net effect depends on the slope of $n(m)$, the (unlensed) galaxy counts as a function of magnitude. A logarithmic slope less than 0.4 (usually the case at visible wavelengths, but barely) will not provide enough "new" sources to overcome the dilution effect, so the source density decreases as κ increases toward the center of a cluster. This *depletion curve* reveals lens parameters, as shown in Fig. 2.5 [88]. Despite the name, the method need not be restricted to one-dimensional information [22]; [88] includes a lens ellipticity and position angle estimate based on a crude depletion map. In practice, measuring magnification is quite difficult, because the slope of $n(m)$ is perilously close to 0.4, and there are few published depletion curve measurements [88,40]. For the remainder of this work, we shall concentrate on algorithms and results using shear, not magnification.

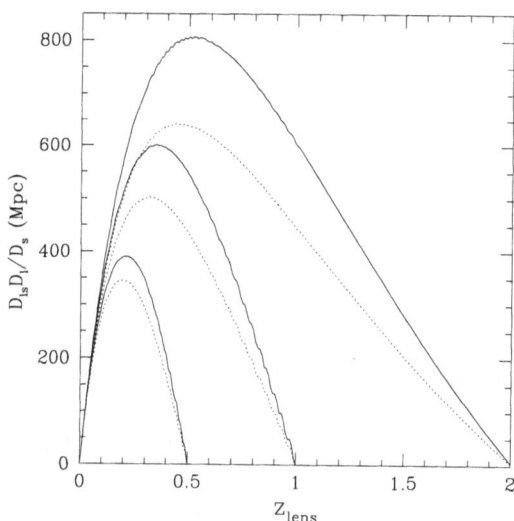

Fig. 2.4. Same as for Fig. 2.3, but as a function of lens redshift, for several values of source redshift (which correspond to the right-hand end of each curve). The lensing efficiency is a very broad function, making it difficult to separate unrelated structures along the line of sight.

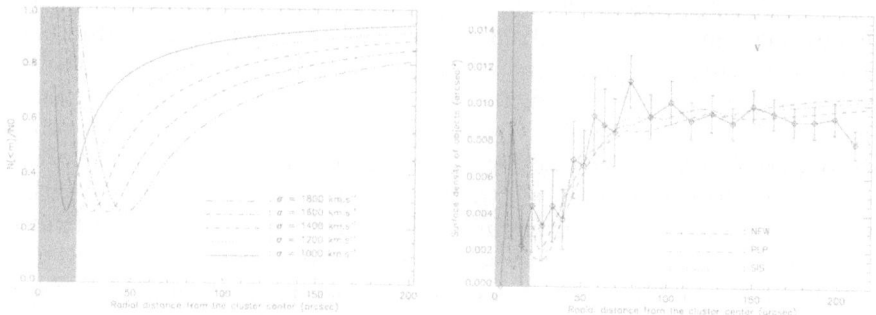

Fig. 2.5. Left: theoretical depletion curves for a variety of lens velocity dispersions (lens mass $\propto \sigma_v^2$). Right: depletion curve observed for MS1008-1224 in V band. From [88].

Angular Resolution. The angular resolution of weak lensing is limited by the areal density of sources. With a shape noise of $\sigma_e \sim 0.3$ and $\sqrt{n}$ statistics, a shear measurement accurate to p percent requires $\sim 1000p^{-2}$ sources. Angular resolution is then set by the area of sky over which these sources are scattered. This in turn depends on the depth and wavelength of the observations; in R there is one source per square arcminute in a one-magnitude wide bin at $R \sim 21.4$, increasing by a factor of ~ 2.5 for every magnitude deeper [122]. A medium deep observation capable of shape measurements to $R \sim 25$ thus yields about

20 galaxies arcmin^{-2} (assuming a bright cutoff $R > 23.5$ to eliminate largely foreground sources), implying that 2 arcmin^2 are required for 5% accuracy in shear.

Getting more sources per unit area requires much more telescope time. Source density will ultimately be limited by confusion — when sources are so numerous that they overlap and hinder shape measurements — around ~ 1000 sources arcmin^{-2} for ground-based data. This implies $\sim 20''$ shear resolution, or better for space-based data, if galaxy counts keep rising at the same rate. However, such depth is hard to come by and must compete against area and wavelength coverage (useful for constraining source redshifts) when planning for a given amount of telescope time.

Another tradeoff commonly used is to sacrifice resolution in one dimension to achieve better resolution in the other. Clusters are commonly analyzed in terms of a radial profile, which assumes they are axisymmetric and allows all sources at a given radius from the cluster center to be averaged together. Less massive clusters and groups can be "stacked" to yield an average profile with reasonable resolution, just as in galaxy-galaxy lensing [57,116].

Source Redshift Distribution. Lack of knowledge of the source redshift distribution is often a limit in calibrating weak lensing measurements. The root of this problem is that deep imaging quickly outruns the ability of even the largest telescopes to provide spectroscopic redshifts for a fair sample of sources.

The recent development of photometric redshift techniques, in which multi-color imaging provides enough spectral information for a reasonable redshift estimate citeConnolly1995,Hogg1998, has brought hope that source redshifts may be estimated to sufficient accuracy from imaging alone. For example, the Hubble Deep Field yielded photometric redshifts accurate to ~ 0.1 per galaxy in the redshift range $0 - 1.4$ with seven filters extending through the near-infrared ($UBVIJHK$) [60]. A look at Fig. 2.3 shows that this provides a reasonable accuracy in distance ratio in most situations. The accuracy improves with the number of filters used, resulting in a tradeoff between accuracy and telescope time. Deep U and infrared imaging are much more expensive than $BVRI$ in terms of telescope time, but it is difficult to effectively cover a large redshift range with only $BVRI$. Few spectral features are to be found in the observed $BVRI$ bandpasses for sources in the redshift range $\sim 1.5 - 3$, which greatly increases uncertainties there.

However, these problems are not fundamental, and photometric redshifts will become routine. They will do much more than help estimate the mean distance ratio required for calibrating lenses. Because sources lie at a range of redshifts, they will provide the opportunity to probe structure along the line of sight (albeit with resolution limited by the width of the lensing kernel). The ultimate goal is *tomography* — building up a three-dimensional view of mass in the universe from a series of two-dimensional views at different redshifts. The combination of weak lensing and photometric redshifts thus promises to be very powerful, but as yet there are not many published examples of combining the two, and little theoretical work on optimal ways of doing so. Although we can expect photometric redshifts to be a routine part of future lensing work, we must be aware of alternative ways of confronting the source redshift problem.

First, some questions can be answered without calibration of source redshifts. The two-dimensional morphology of a cluster lens is one example — the source redshift distribution should not depend on position (as long as magnification is negligible and cluster members do not contaminate the source sample). Similarly, source redshifts are not required for discovery of mass concentrations in surveys, but without them, the volume probed is unknown. Clearly, the questions which can be answered this way are limited.

A more general calibration strategy is through additional, identical observations of a "control lens" of known redshift and mass (e.g. a cluster with a dynamical, X-ray, and/or strong lensing mass estimate). This does allow estimation of the mean distance ratio to a population of sources much too faint to reach with spectroscopy, but it certainly has its limits. It is difficult to obtain identical observations, and the (probably considerable) uncertainty in the mass of the control lens becomes a systematic for the rest of the data. But more fundamentally, shear from the control lens samples only that part of the source distribution which is behind the control lens, so that strictly speaking, a control lens must be at the same redshift as the target. For weak lensing by large-scale structure, the distribution, not simply the mean distance ratio, is required. This would require control lenses at a range of redshifts, which is impractical. Photometric redshifts should do a much better job with more realistic data requirements. Even in the age of photometric redshifts, though, this method will have its role. The shear induced by calibrated lenses will provide a check on photometric redshift estimates, which may not be checkable with spectroscopy if applied to very faint sources.

Another strategy is keeping the imaging as shallow as current redshift surveys, which go to $R \sim 24$. One can then look up the median redshift for any magnitude cut; for $23 < R < 24$, for example, the median redshift is 0.8 [29]. Even the redshift distribution is known to some extent, with 120 sources in that magnitude slice in the survey cited. Shallow imaging need not probe a small volume, as a large area can be covered with a reasonable amount of telescope time. But it does limit the distance probed and the angular resolution of the mass reconstruction (because the areal density of sources is low at $R \leq 24$). This strategy also limits selection of sources based on color, which is very useful for limiting contamination by galaxies in a cluster being studied (or for de-emphasizing foreground contamination in general) because the median redshift of a color-selected sample is not yet something that can be looked up in a redshift survey.

Intrinsic Alignments. The crucial assumption in weak lensing is that the sources have random intrinsic orientations, so that any departure from randomness is due to lensing. This assumption is worth examining before proceeding further. We will concentrate on potential damage to measurements of weak lensing by large-scale structure (cosmic shear), because the lensing signal from clusters is usually at a much higher level. However, it is worth keeping in mind that all applications of weak lensing could be affected at some level.

The first detections of cosmic shear in 2000 motivated several analytic [25] and computational [54,36] studies of intrinsic alignment mechanisms, and the field is still sorting itself out. There are several mechanisms which could produce such intrinsic alignments, including tidal stretching of galaxies in a gravitational potential, and coupling of the potential to the spin vectors of galaxies [34]. The amount of alignment predicted as a function of angular scale varies greatly depending on the mechanism and the strength of the coupling; it remains unknown which model, if any, is correct. However, in most scenarios, intrinsic alignments would represent a $\lesssim 10\%$ contamination of the cosmic shear measurements.

While the situation is still evolving, one rule is certain: the effect of any intrinsic alignment is diluted when sources lie at a large range of redshifts, as is naturally the case in deep imaging. As we shall see, the signal from lensing by large-scale structure increases with source redshift. Hence, lensing must dominate at high enough source redshift, and intrinsic alignments at low enough source redshift. This is illustrated by Fig. 2.6, which shows predicted intrinsic alignment and cosmic shear levels for several source redshifts. For shallow surveys like the Sloan Digital Sky Survey [141] (SDSS), intrinsic alignment may frustrate attempts to measure cosmic shear, but deeper surveys specifically designed to measure cosmic shear are safe. Indeed, the roughly one million spectroscopic redshifts SDSS plans to acquire will be invaluable in measuring intrinsic alignments precisely, and their measurements, after scaling to higher redshifts, in turn may facilitate estimation and even removal of intrinsic alignment effects from the deeper surveys. Deep surveys may also be able to provide a lensing signal using only sources which cannot be physically associated, as indicated by their photometric redshifts. Density reconstruction methods in the presence of intrinsic alignments are already being investigated [83].

The two or three detections of intrinsic alignments in real data are indeed at low redshift. Ellipticity correlations have been reported in SuperCOSMOS data [23], but because there is no redshift information, intrinsic alignments can only be inferred (the median source redshift is estimated to be < 0.1, so that the inferred cause is intrinsic alignments rather than lensing). Spin alignments have been found in the Tully catalog, which consists of several thousand nearby (within a few Mpc) spirals [102], and in the PSCz, a redshift survey of 15500 galaxies detected by the IRAS infrared satellite mission [84]. Because these catalogs have redshift information, these represent solid detections. However, spin correlations are one step removed from ellipticity correlations, which are the relevant quantity for lensing.

It may also be possible to extract intrinsic alignments from the lensing data itself. To first order, lensing produces a curl-free, or E-type (in analogy with electromagnetism) shear field. (Multiple scattering can produce a weak divergence-free, or B-type field, but that can be safely ignored for the moment.) Therefore, decomposition of a measured shear field into E-type and B-type fields might allow separation of the lensing and intrinsic alignment effects [35]. This decomposition

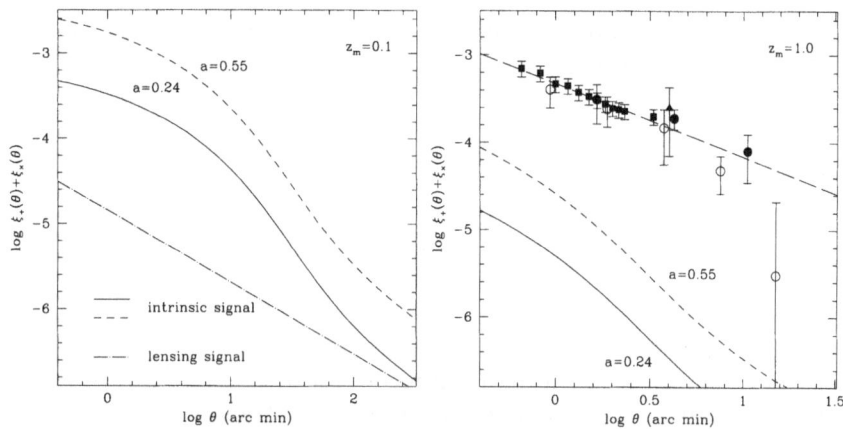

Fig. 2.6. The importance of intrinsic alignments depends strongly on source redshift. The expected levels of ellipticity correlation (defined in Sect. 2.3.1) due to intrinsic alignments and to weak lensing by large-scale structure are shown for low-redshift (median source redshift $z_m = 0.1$) surveys in the left panel and for high-redshift ($z_m = 1$) surveys in the right panel. In each case, the straight line indicates the expected signal from weak lensing, and the curves indicate the expected signal from intrinsic alignments, for two different values of a spin-coupling parameter, giving some idea of the modeling uncertainties. The right panel also contains cosmic shear measurements from the literature, which all happen to have $z_m \sim 1$. Adapted from [34].

is difficult, but a crude indicator of the B-type field is the traditional 45° test. In this test, a lensing signal should vanish when one component of the shear is exchanged for the other (equivalent to rotating each source by 45°), and nonzero results would indicate a systematic error. All published cosmic shear results were vetted using this test among others, with no indication of contamination. However, not all intrinsic alignment mechanisms produce B-type power; an example is tidal stretching (tidal fields are the basic mechanism for both stretching and lensing, after all). Based on other astrophysical arguments, tidal stretching is not likely to be significant [35], but even angular momentum coupling models can produce much more E-type than B-type correlations [86]. Passing the 45° test is a necessary but not sufficient condition for confidence in the results.

In summary, intrinsic alignments are not to be dismissed. They must be addressed and may even dominate the lensing signal in certain low-redshift applications. However, the dilution effect of a broad source redshift distribution means that none of the conclusions of weak lensing to this point can be called into doubt. Ongoing and future weak lensing studies may have to apply small corrections for this effect, but how small is still uncertain. Accurate corrections will probably be available by the completion of the SDSS, which will do much to increase our knowledge of intrinsic alignments in the nearby universe.

2.1.5   Measuring Shear

In most weak lensing work, a source galaxy is approximated as an ellipse fully described by its quadrupole moments

$$
I_{xx} \equiv \frac{\Sigma I w x^2}{\Sigma I w}
$$
$$
I_{yy} \equiv \frac{\Sigma I w y^2}{\Sigma I w} \qquad\qquad (2.3)
$$
$$
I_{xy} \equiv \frac{\Sigma I w x y}{\Sigma I w}
$$

where $I(x, y)$ is the intensity distribution above the night sky level, $w(x, y)$ is a weight function, the sum is over a contiguous set of pixels defined as belonging to the galaxy, and the coordinate system has been translated so that the first moments vanish (*i.e.* the centroid of the galaxy is chosen to be the origin of the coordinate system). Early work used intensity-weighted moments ($w = 1$), but it was realized that this produces ellipticity measurements with noise properties that are far from optimal or even divergent. Now, w is usually chosen to be a circular [73] or elliptical [15] Gaussian, which deweights the outer pixels which have a big lever arm but low signal-to-noise. The two ellipticity components can be defined as [130]

$$
e_+ = \frac{I_{xx} - I_{yy}}{I_{xx} + I_{yy}}
$$
$$
e_\times = \frac{2 I_{xy}}{I_{xx} + I_{yy}} \qquad\qquad (2.4)
$$

These are related to the scalar ellipticity ϵ and position angle ϕ by

$$
\epsilon = (e_+^2 + e_\times^2)^{\frac{1}{2}}
$$
$$
\phi = \frac{1}{2} tan^{-1}\left(\frac{e_\times}{e_+}\right) \qquad\qquad (2.5)
$$

Then a simple estimate of the shear in the weak lensing limit is $\gamma_i = \langle e_i \rangle / 2$, where the brackets denote averaging over many sources (perhaps with weighting of the sources based on estimated measurement errors, redshift, etc.) to beat down shape noise. Note that this definition of ellipticity differs from that in (2.2) by a factor of two; both definitions are presented here because both are common in the literature. This latter estimator sometimes called the *distortion* statistic. Also, there are alternative formulations in terms of octupole moments [48], Laguerre expansions [15] and shapelets [107,108,27]. Before applying any of these estimators, we must account for the effects of point-spread function (PSF) anisotropy and broadening.

PSF Anisotropy. No optical system is perfect, and PSFs on real telescopes tend to be $\sim 1 - 10\%$ elliptical. This constitutes a huge systematic error, of the order of the shear induced by even a massive cluster, and it must be removed as completely as possible before analyzing any galaxy shapes, and monitored afterward. The removal can be done *after* measuring shapes, by essentially subtracting the moments of the PSF from the galaxy moments, but a more computationally stable method is to remove this effect from the image, *before* measuring shapes. This is done by convolving the image with a kernel with ellipticity components opposite to that of the PSF [47]. The raw PSF is almost certainly position-dependent; therefore the circularizing kernel is also, but the convolved PSF is everywhere round. A round PSF is called *isotropic*, but keep in mind that this does not imply homogeneous: The convolved PSF may vary somewhat in size, because of the position dependence of the original PSF and of the small broadening introduced by the kernel. A more sophisticated scheme would introduce more broadening in the right places, leading to a PSF which is homogeneous as well as isotropic. It is also possible to choose a sharpening kernel, but this would amplify the noise in the image.

Figs. 2.7 and 2.8 illustrate the effectiveness of the convolution procedure. Although there are low-level residuals in the convolved image, their lack of spatial correlation means that they will have difficulty masquerading as a weak lensing effect. Note that these PSF anisotropies change with time, as telescope temperature, focus, and guiding drift, so that each exposure must be treated separately. A possible benefit here is that if the anisotropies are really uncorrelated temporally, coaddition of multiple exposures will beat down the shape errors. Also, each CCD in a mosaic must be treated separately, as some discontinuities may arise from small differences in piston between devices.

PSF Broadening. Any effect which broadens the PSF will reduce the measured ellipticities of source galaxies which are not much larger than the PSF—generally including the distant galaxies most appropriate for lensing—because they will be broadened relatively more along their minor axes than along their major axes. In ground-based data, the dominant effect is "seeing", the broadening of the point-spread function due to turbulence in the atmosphere. Note that this is distinct from PSF anisotropy, which is caused by the telescope and camera optics. In fact, seeing produces a circular PSF as long as the integration time is much longer than the coherence time of the atmosphere (very roughly 30 ms at visible wavelengths); anyone who has observed in terrible seeing has probably noticed that at least the PSF is nicely round ! The effects of PSF anisotropy and broadening are sometimes called "shearing" and "smearing", respectively. The former has the effect of introducing a spurious weak lensing signal if uncorrected, while the latter has the effect of reducing any weak lensing signal.

There are several ways of correcting for smearing. The first is measuring the dilution of a simulated weak lensing signal relative to an unsmeared image, either simulated or perhaps from the Hubble Deep Fields. The seeing-free images are sheared by a known amount, convolved with the point-spread function of the real data, repixelized, and the shear measured. This has the advantage of including some effects which cannot be accounted for analytically, such as the coalescing of separate objects into an apparently elliptical single object.

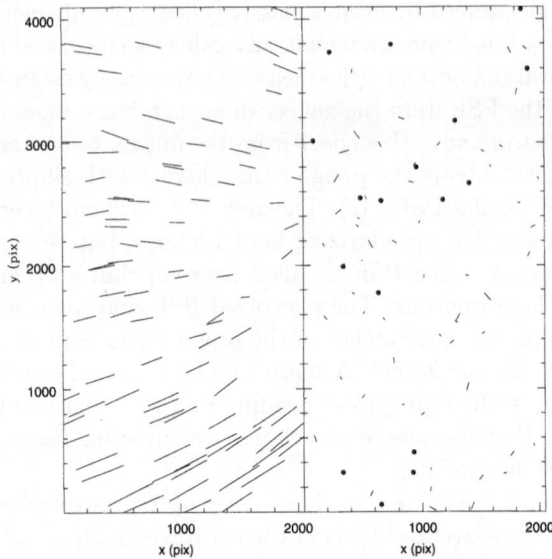

Fig. 2.7. Point-spread function correction in one 2k×4k CCD. Shapes of stars, which as point sources should be perfectly round, are represented as sticks encoding ellipticity and position angle. Left panel: raw data with spatially varying PSF ellipticities up to 10%. Right panel: after convolution with a spatially varying asymmetric kernel, ellipticities are vastly reduced (stars with $\epsilon < 0.5\%$ are shown as dots), and the residuals are not spatially correlated as a lensing signal would be.

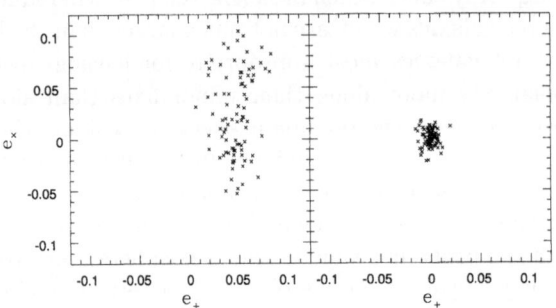

Fig. 2.8. Another way of plotting the efficacy of the PSF correction, often seen in the literature. For the same dataset shown in Fig. 2.7, the ellipticity components of point sources are shown in a scatterplot, before and after correction. This type of plot hides any spatial correlation which may exist among the residuals, but Fig. 2.7 shows that the residuals are uncorrelated in this case.

On the other hand, a global correction is rejected by those who prefer to tailor the corrections to individual sources; after all, a large galaxy is smeared relatively less by seeing than is a small one. The advocates of this approach tend to use the KSB method, an analytical approach which takes into account the size of the PSF and of each source [73]. The KSB method also accounts for PSF shearing, but it can just as well be applied to a convolved image. See [70,71,82,15] for limitations of and possible successors to the KSB method. These approaches also weight each source according to its ellipticity uncertainty when computing a shear estimate.

Source Selection. Not every source in a deep image should be included in a shear measurement. A typical deep image includes stars, other unresolved sources, foreground galaxies, cluster members if the target lens is a cluster, and spurious objects, such as bits of scattered light around very bright stars. Getting rid of these unwanted sources is something of an art, which must reflect the particular data set, but generally there are four kinds of cuts. Magnitude cuts help get rid of stars (for reasonable galactic latitude, stars outnumber galaxies for $R \lesssim 22$ while galaxies greatly outnumber stars for $R > 23$) and bright foreground galaxies. Galaxies have a broad luminosity function, so such a cut is never completely effective at eliminating the foreground, but it helps. Color cuts seek to emphasize the faint blue galaxies at $z \sim 1$ [124]. If the target is a cluster, the cut should be blueward of the cluster's color-magnitude ridge. Even so, some cluster members and other foreground galaxies will survive. Size cuts eliminate unresolved objects, which at the relevant magnitudes include some stars, but mostly unresolved galaxies. Finally, cuts designed to insure that an object is not spurious must depend on the type of data available. Examples include rejecting objects that appear on only one of a multicolor set of images, and rejecting high ellipticity objects which are likely to be unsplit superpositions of two different objects.

Sanity Checks. There are a number of sanity checks that should be performed before believing any weak lensing result. In addition to the 45° test mentioned above, randomizing source positions while retaining their shapes should result in zero signal. Another good sanity check is correlating the source shapes with an unlensed control population, such as a set of stars. Finally, there are checks on the basic integrity of the catalog, such as the position angle distribution of sources, which might reveal spurious objects aligned with the detector axes. Because setting the source selection criteria can be somewhat subjective, it is also good to check that the results do not depend crucially on the exact magnitude or color cut.

2.2   Lensing by Clusters and Groups

Clusters of galaxies have long been studied from two somewhat opposing points of view. Visible from great distances, they are a convenient tracer of structure in the universe back to roughly half its present age. When examined individually, they are interesting astrophysical laboratories in their own right, with a variety of physical conditions and histories. But if so, they cannot be simple cosmological probes. So the study of clusters as astrophysical laboratories must inform and refine the study of clusters as cosmological probes.

What lensing adds to the study of clusters is a direct mass measurement without any assumptions about the dynamical state of the cluster. The first clusters were "weighed" in the 1930's with the dynamical method—assuming that clusters are in virial equilibrium, the virialized mass is easily computed from the velocity dispersion. In the late 1980's, X-ray imaging of hot intracluster gas began to provide mass estimates, assuming hydrostatic equilibrium. In the 1990's, lensing began to provide mass estimates free of any such assumptions. The frequent agreement of the three types of estimate indicates that the dynamical assumptions are often valid, but the exceptions need to be identified. Those exceptions must be discarded from any samples used as cosmological probes, but they are often studied more closely for what they might reveal about mergers or other nonequilibrium processes.

In the past decade, it was enough simply to compare lensing measurements of cluster masses with those provided by other techniques. Driven by advances in wide-field detectors, we can now use lensing to *search* for clusters (or at least mass concentrations), and even estimate their redshifts. Shear-selected samples of clusters, free of any bias toward baryons that optically and X-ray selected samples might have, are currently being compiled. Comparison of the different types of samples will be instructive, either by confirming the use of traditional baryon-selected samples as cosmic probes, or perhaps by providing some counterexamples.

2.2.1   Masses and Profiles

The first evidence of lensing by clusters came in the late 1980's in the form of strongly lensed giant arcs [85,118], which were used to constrain the mass inside the radius at which the arcs appeared. This was soon extended to somewhat less strongly lensed "arclets", and by 1990, to the first detection of what we now call weak lensing, the coherent alignment of thousands of weakly lensed background galaxies [125]. This alignment was measured in terms of the *tangential shear* γ_t, which is the component of shear directed tangential to an imaginary circle centered on the cluster and running through the source. The tangential ellipticity of a source is $e_t \equiv \epsilon \cos(2\theta)$ where θ is the angle from the tangent to the major axis of the source (Fig. 2.9). When computing e_t from the ellipticity components, use the rotation

$$e_t = +e_+ \cos(2\beta) + e_\times \sin(2\beta)$$
$$e_c = -e_+ \sin(2\beta) + e_\times \cos(2\beta) \tag{2.6}$$

where the angle β is also shown in Fig. 2.9. Here e_c is a control statistic measuring the alignment along an axis 45 degrees from the tangent, which is not affected by an axisymmetric lens.

Methods for constraining cluster masses using tangential shear followed soon after the first detection of the effect [91]. The most important of these is *aperture*

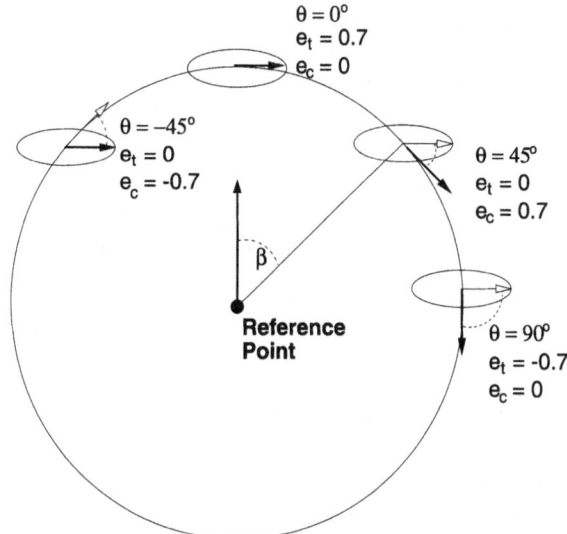

Fig. 2.9. Tangential ellipticity e_t of an $\epsilon = 0.7$ source with respect to a reference point. e_t carries the lensing signal, and the 45-degree component e_c serves as a control. In the presence of lensing but not shape noise, all sources would have $e_t > 0$ and $e_c = 0$; in practice $\langle e_t \rangle > 0$ and $\langle e_c \rangle = 0$.

densitometry, which relates γ_t to the difference between the mean projected mass density inside a radius r_1 and that between r_1 and a larger radius r_2 [45]:

$$\bar{\kappa}(<r) - \bar{\kappa}(r_1 < r < r_2) = \frac{2}{1 - r_1^2/r_2^2} \int_{r_1}^{r_2} \frac{\gamma_t}{1 - \kappa(r)} d\ln r. \qquad (2.7)$$

The factor $1-\kappa$ can be ignored in the weak lensing limit, but for massive clusters may not be ignored, leading to an iterative solution for κ (*e.g.* [47]). The left hand side of this equation is sometimes called the *zeta statistic* $\zeta(r_1, r_2)$. Note that this formula makes the mass sheet degeneracy explicit by specifying only relative values of κ; the best that can be done is extend r_2 to a very large value, at which κ should vanish.

A profile can be built up by repeatedly applying this statistic at a sequence of different r_1. However, this makes the points in the profile dependent on each other, as they use much the same data. If the goal is to find the best fit of a given type of profile, it is simpler to compute γ_t in a series of independent annuli and fit the shear profile expected from the mass model straightforwardly with least-squares fitting.

Weak lensing mass profiles are usually well fit by a singular isothermal sphere (SIS) or Navarro-Frenk-White (NFW, [98,99]) profile (see [76] for an extensive list of profiles used in lensing, along with their associated formulae). However, the nature of weak lensing makes it difficult to distinguish between models on two accounts. First, shear profiles do not have good dynamic range because the uncertainty in shear measurements increases dramatically at small radii, where there are not enough sources to beat down the shape noise. This is illustrated in the top panel of Fig. 2.10. Note that this figure is for a very massive cluster; the signal-to-noise ratio can only be lower for less-massive clusters. Second, mass profiles which differ significantly *inside* the radius where shear is measured can produce shear profiles which differ significantly *only outside* that radius. This is illustrated in the middle panel of Fig. 2.10. The ability to distinguish between NFW and SIS (or more generally, power-law profiles) thus depends strongly on the size of the field [77], but Fig. 2.10 demonstrates that a significant ambiguity remains even with a state-of-the-art imager with a 35' field. Weak lensing is therefore not definitively revealing cluster mass profiles, as one might have expected. Progress toward larger fields will be slow, as most large telescopes already have imagers which fill their usable fields of view. More likely, progress will come by adding magnification information. Finally, note that the most active (and revealing of the nature of dark matter) controversy surrounding cluster profiles involves cuspiness at the center, and this is not well addressed by weak lensing, with its poor angular resolution.

Initially, weak lensing analyses of clusters concentrated exclusively on the most massive clusters which were guaranteed to give a good signal. As the technique has matured, it has been extended to less massive, but more typical clusters [139]. The ultimate extension has been to groups; although a group by itself does not provide enough shear to get an accurate mass estimate, they can be "stacked" as in galaxy-galaxy lensing to build up an average profile with reasonable signal-to-noise [57]. The idea of stacking to obtain a good estimate of the average profile has been used for typical clusters as well [116]. Caution is required when interpreting "average" results, though, because they may be biased by a few unrepresentative systems, or in the worst case, meaningless if the sample is sufficiently heterogeneous.

Mass estimates of clusters and groups derived from weak lensing generally agree with estimates from velocity dispersions and X-ray imaging (see [89] for a list of published mass estimates as of 1999; there are now too many to list). At one time there was an apparent systematic discrepancy between (strong) lensing and X-ray estimates [93], but it was shown to be due to the complexities of the X-ray-emitting gas dynamics [3]. Hydrostatic equilibrium alone was shown to be less constraining than initially thought; temperature maps were needed [8]. XMM and Chandra now provide these, along with vastly improved angular resolution which allows for better treatment of cooling flows, and the first results show good agreement with lensing [4]. Of course, not every cluster behaves so well, and when there is disagreement a closer look often reveals interesting astrophysics such as cooling flows, mergers and their associated "cold fronts" or shock heating. In fact, ∼50% of clusters show some substructure in the X-ray [114]. Weak lensing can still supply the total mass, but due to its poor angular and line-of-sight resolution, the detailed work of disentangling the structures must be left to X-ray and dynamical measurements.

The three approaches, after all, have some fundamental differences which are not often mentioned. Dynamical estimates based on the virial theorem measure

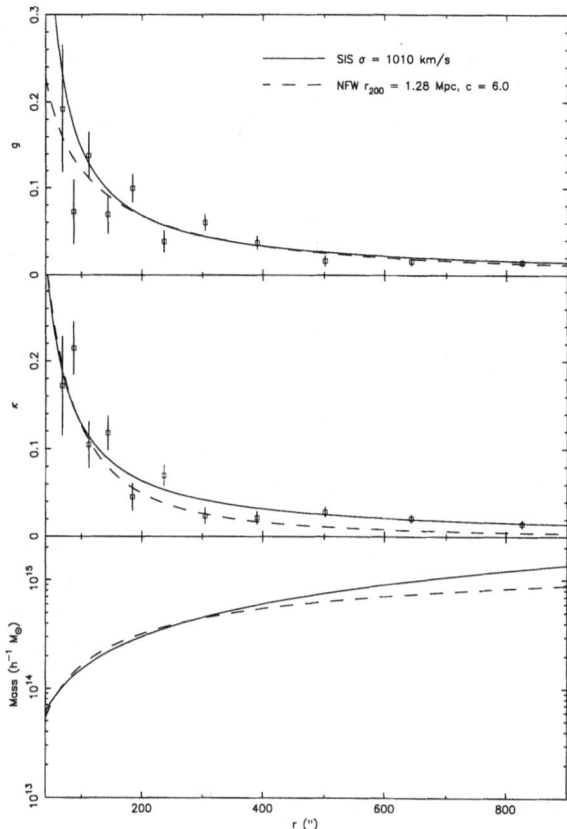

Fig. 2.10. Comparison of shear, convergence, and mass profiles for the massive cluster Abell 1689. *Top:* (Reduced) shear profile, with best-fit SIS and NFW profiles almost indistinguishable. *Middle:* κ profiles, showing that the NFW model falls off much more steeply than the SIS at large radii. This could barely be seen in the shear profile because of the nonlocality of shear. Despite appearances, the data do not favor the SIS. The points here are plotted assuming that outside the largest radius measured, the shear falls off as an SIS. If NFW is instead assumed for radii outside the measurement area, the points at large radius fall significantly. *Bottom:* Enclosed mass profiles of the two models. The physical scale is roughly 2 kpc arcsec^{-1}. From [28].

the total virialized mass of the cluster, while lensing can only measure mass projected inside a certain radius. Even with a simplifying assumption such as spherical symmetry, a fair comparison of the two is difficult. Virial masses go as the square of the velocity dispersion, so small-number statistics and outliers can have a large effect on the mass [109]. Mass estimates from a dozen members may be good for a back-of-the envelope comparison, but beyond that should be treated with extreme caution. Velocity *profiles* would be more comparable to lensing and X-ray data. These are available for few clusters, but thanks to multi-object spectrographs on large telescopes, such detailed dynamical analyses are becoming more common [16]. X-ray emission is proportional to the square of the density, so it is more sensitive to substructure than is lensing, which is simply proportional to the density. A fourth approach, the Sunyaev-Zel'dovich effect (SZE), measures the decrement in the cosmic microwave background (CMB) caused by upscattering of CMB photons by the hot intracluster gas. Like lensing, it is proportional to the density, but like X-ray emission, it depends on the density of baryons, rather than all matter. The first SZE measurements are starting to arrive and will soon offer their unique point of view.

Lensing stands out from X-ray and dynamical methods in being a projected statistic, so it is worth asking whether this introduces any bias. It appears that anisotropy in simulated clusters has little systematic effect [21], and so do uncorrelated structures along the line of sight [56]; both effects are around the 5% level. However, in reality there are also correlated structures along the line of sight, and these can bias masses upwards by tens of percent [26,90]. There is general agreement that the effect of other structures along the line of sight increases with aperture size. The bias changes with redshift in two ways. First, it is minimized when the cluster is near the peak of the lensing kernel (Fig. 2.4), because other structures will be deemphasized. Second, younger clusters may have more nearby material, although this effect has not been investigated thoroughly [90]. Finally, note that the mass function is susceptible to bias even when an estimator is unbiased but has scatter, because there are more low-mass clusters to be scattered up than high-mass clusters to be scattered down (this applies equally to other types of mass estimates such as dynamical and X-ray) [90].

2.2.2 Two-Dimensional Structure

From the first detection of weak lensing, it was realized that the tangential shear procedure could be repeated about any reference point, not only the cluster center. By repeating it at a grid of points, a two-dimensional "mass map" (actually a map of κ) was constructed [125]. This was soon put on a firm theoretical footing by the derivation of a relationship between the Fourier transforms of κ and γ, starting from their relationship as different linear combinations of the second derivatives of the lensing potential ψ [72] (see Chap. 1 for the relationship between ψ and κ). Essentially, κ can be expressed as a convolution of γ over the entire plane (there is also a real-space equivalent [47]). Of course, observations do not cover the entire plane, a problem called the *finite field effect* [10]. This is another manifestation of mass sheet degeneracy, as the spatial variation, but not the mean value, of κ can be reconstructed. Several reconstruction methods based on magnifications have been proposed to combat this problem [22,11], but, as mentioned above, magnification is very difficult to measure, and these methods have not been widely used. To a large extent, technology has solved the problem, at least for clusters, by providing ever-wider fields of view, at the edges of which κ is presumably negligible.

In addition to direct *reconstruction* methods, there are *inversion* methods which solve for ψ, from which κ can be derived [12]. An extensive comparison of different methods found none to be clearly superior [119], although inversion methods tend to make it easier to include additional constraints such as those from magnification measurements or strong lensing features.

Many clusters have now been mapped using these techniques, and the mass distributions recovered are generally not surprising. That is, they roughly follow the optical and X-ray light distributions, on the scales which weak lensing is able to resolve. A vivid example of two-dimensional mass reconstruction is that of the supercluster MS0302+17 [74] (Fig. 2.11). This supercluster contains three clusters separated by 15-20′ on the sky ($\sim 3 - 4$ Mpc transverse separation at a redshift of 0.42). All three clusters are recovered by the reconstruction algorithm,

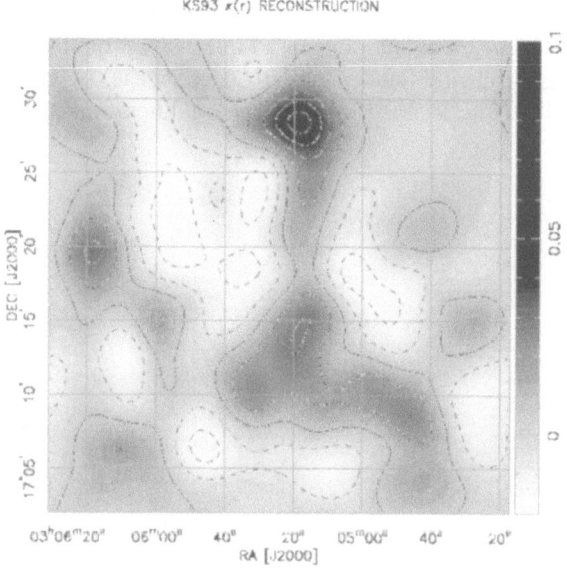

Fig. 2.11. Mass map of the supercluster MS0302+17, smoothed on a scale of 90 arcseconds (black indicates higher density). Each of the three densest blobs corresponds to a known galaxy cluster. From [74].

above the level of other, presumably noise, peaks. This result is robust: it remains when the source selection criteria are varied, and it disappears when the source positions are randomized. There appears to be a filament connecting two of the clusters, but the authors advise caution, as the signal-to-noise is low, and real filaments are not expected to have much contrast against all the other filaments and sheets expected to lie between sources and observer. An equally striking reconstruction of the Abell 901/902 supercluster was recently published [50]. The close correspondence of mass peaks and known clusters says something about the predictability of dark matter, as discussed below.

Despite these successes, it is worth remembering that a map (or radial profile) of κ is not a map of mass. κ can be converted to mass only with a careful calibration of the source redshifts, which must include an estimate of source contamination by the cluster itself. In massive clusters, magnification provides another source of error by increasing the mean redshift of sources which have been selected according to an apparent magnitude cut. This results in Σ_{crit} being a function of radius, as it decreases at small radii where the sources of a fixed magnitude tend to be more distant [47]. While much attention has been paid to optimizing the formal reconstruction methods, these more mundane problems require equal attention.

2.2.3 Mass and Light

Does light trace mass ? The answer must be at least a qualified yes, because the projected shapes of cluster lenses tend to agree with the shapes suggested by their emitted light (Fig. 2.12). However, the qualifications are important !

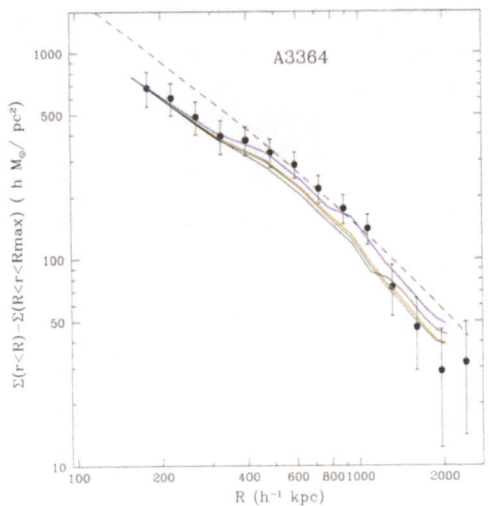

Fig. 2.12. Projected mass and light density profiles of Abell 3364. The light profiles were observed in observer-frame B_j (blue line), V (green), R (red), and I (black) filters, and computed in the same differential apertures used for the mass. The light profiles have each been shifted vertically to intersect the innermost mass point, hence they are in arbitrary units. Mass follows light surprisingly well on all measurable scales. The dotted line shows the shape of an isothermal profile, which is not quite a straight line with this estimator, to guide the eye (it has not been fit to the data). The two lowest mass points are approaching the level of systematic error estimated from the point-spread function. Note that in the aperture densitometry method, error bars on adjacent points are not independent, so that the errors should be thought of as a band. From [139] (see Color Plate).

First, the correspondence holds only on scales larger than galaxies. The vast majority of visible-wavelength light from clusters comes from individual galaxies, not diffuse emission. Although weak lensing is not well suited to examine small scales, there is ample evidence from strong lensing that cluster mass distributions (like their X-ray emissions) do not peak on galaxy scales [123].

Second, not all light is equal. Blue light is dominated by very young stars, while established stellar populations which presumably trace mass better tend to be red. Hence small variations in star formation could scatter the ratio of mass to blue light M/L_B widely from system to system, but M/L_R should be much more stable. Unfortunately, the literature has a tradition of quoting M/L_B, which obfuscates the issue of whether mass is traced by light from established stellar populations. Compounding this confusion are different methods of computing rest-frame emission given the observed emission, and the occasional quotation of M/L at $z = 0$, meaning that L has first been adjusted to a value that it would have at $z = 0$ given passive evolution. Because stellar populations fade with time, L can decrease, and M/L increase, significantly from its *in situ* value.

Third, even if M/L were more consistently defined, it may not be constant spatially or as a function of scale. Although the literature is full of mass reconstructions which generally follow smoothed light distributions, the M/L found varies widely, from $\sim 80h$ to $\sim 800h$. Some of this is no doubt due to different methods mentioned above, but there is reason to believe that not all of it is. For example, the high value of $\sim 800h$ for MS1224+20 was found independently by two different investigators [45,46]. Also, attempts to uniformly treat samples of ~ 10 clusters have found a range of M/L within the samples [117,139]. There are some hints that some of the scatter may be due to a trend of increasing M/L with cluster mass.

If so, this follows a broader trend in which groups have lower M/L than the typical cluster [57], and typical galaxies have still lower M/L. The idea that M/L might depend on environment is called *bias*. Specifically, bias is when light is more concentrated than mass; the reverse, *antibias*, may also occur. However, indications of antibias can be understood as a simple consequence of stellar evolution and the choice of a blue bandpass [7]. In the end, M/L may be more a question of star formation history and bandpasses than of the nature of dark matter.

An alternative approach in terms of galaxy-mass correlations may offer more promise. Because large mass concentrations are clearly more associated with early-type galaxies than with later types (the *morphology-density relation*), restricting the analysis to early-type galaxies might reveal a tighter relationship to mass. This was first done for the MS 0302 supercluster, shown in Fig. 2.11 [74]. A cross-correlation between the projected mass density and that predicted from the early-type galaxies revealed a strong relationship, which did not vary with density, whereas a simple M/L would have acquired variations from the variations in star formation activity. This approach has been extended to the field, with similar conclusions [136]. Correlations between mass and light were found to be not so simple in the Abell 901/902 supercluster [50], but perhaps the difference is due to all light, not only that from early-type galaxies, being used in the Abell 901/902 work. Still, the correlation of mass with early-type galaxies must fail on some smaller scale, as we know that galaxy groups have mass but generally no early-type galaxies. Clearly more work is required in this area, as the correlation of mass with different types of emitters may provide clues to the nature of dark matter.

2.2.4 Clusters as Cosmological Probes

There is a hidden agenda behind all the effort that has gone into measuring cluster M/L: if cluster M/L is representative of the universe in general, the mean density of the universe Ω_m can be estimated simply by scaling the local luminosity density by this ratio. This is one version of the *fair sample hypothesis*, and it is one of the many ways to use clusters as cosmological probes. These can be divided roughly into methods that extrapolate from physical conditions in clusters (using clusters only because they facilitate certain measurements), and methods which use clusters to diagnose formation of structure in the universe.

Scaling the luminosity density by M/L is the simplest example of the first type of method. The fair sample hypothesis remains just a hypothesis, but it has nevertheless spawned many estimates of Ω_m, which tend to be $\lesssim 0.4$ [117,74,139,57]. However, the apparent variation of M/L with environment and age makes this approach suspect, and it is worth asking whether any property of clusters other than light could be used in a similar scaling argument.

The best candidate is the baryon fraction f_b, the ratio of baryonic mass to total mass. Because there is no reason to believe that infall into clusters favors baryons over dark matter or vice versa, f_b is plausibly equal to Ω_b/Ω_m. In addition to being plausible, the baryonic hypothesis is easier to investigate with simulations of structure formation, because tracking the baryons is easy in such simulations; the hard part is simulating star formation and the resultant light emission. With Ω_b fairly well known from Big Bang nucleosynthesis arguments [24], a determination of f_b would quickly yield Ω_m. Lensing can provide an estimate of the total mass, while SZE measurements can probe the dominant baryonic component, the intracluster gas. Simulations indicate that the combination should reveal f_b to 10% or better [142,39]. The first results from real clusters (but with total mass estimated from X-ray emission rather than lensing) indicate $\Omega_m \sim 0.25$ [51]. It should be noted that any census of baryons is likely to be incomplete, as they can take many forms which are difficult to detect (brown dwarfs, planets, etc.). Hence this method provides a lower limit to f_b and an upper limit to Ω_m.

There is always a chance that physical properties of clusters such as f_b are simply not representative of the universe in general. A second and more powerful class of cosmological probe uses clusters as tracers of structure. Only their mass is important, and in particular, their *mass function*, the number density of clusters as a function of mass. The redshift evolution of the cluster mass function is a probe of Ω_m: all else being equal, a high-density universe should show more recent evolution than a low-density universe. In fact, it has been argued that the existence of even one massive cluster at high redshift (*e.g.* MS1054 at $z = 0.83$ [38] and now also ClJ1226.9+3332 at $z = 0.88$ [41]) demonstrates that evolution has not been as rapid as required if $\Omega_m = 1$ [19]. Conclusions based on a few massive clusters are suspect, however, because as the extreme tail of a distribution, their numbers are highly dependent on the assumption of Gaussianity in the primordial fluctuations. The argument can even be turned around: given an independent measure of Ω_m, cluster counts can put strong constraints on primordial non-Gaussianity [106,78]. With plentiful wide-field data now available and with weak lensing techniques having been honed on less massive clusters, it will soon be possible to construct an honest mass function, which will constrain both quantities [81]. The redshift evolution of the cluster mass function can also constrain dark energy [55,67]. Without any uniform weak-lensing cluster samples, though, we must defer this discussion to Future Prospects and turn our attention to progress in obtaining such a sample.

2.2.5   Shear-Selected Clusters

The use of clusters as cosmological probes centers on clusters as mass concentrations, not as collections of galaxies and gas. Yet all cluster samples compiled to date have been based on emitted light from galaxies (*e.g.* [2]) or from a hot intracluster medium [20]. Because these mechanisms do not involve dark matter, which is the dominant component by mass, a mass function based on these samples may well be biased. In addition, the r^{-2} falloff of emitted light implies that the high-redshift end of such samples will always be dominated by the most luminous clusters, which potentially introduces another bias. Shear-selected clusters are needed to investigate these potential biases and provide a clean mass function, and this is currently an active area in weak lensing.

First, a note on terminology. "Cluster" implies a collection of galaxies, but if a large mass concentration with no visible galaxies were to be identified, it would probably be called a "dark cluster". Although "dark matter halo" would be a more accurate term, it is used almost exclusively in theoretical and computational papers, not observational work. Here we shall continue to use the term "cluster", but we emphasize that this is a working hypothesis. After large samples of shear-selected mass concentrations are thoroughly followed up with other methods, it will become clear if a different term is more appropriate.

Unfortunately, such samples are not available yet. Although many previously known clusters were studied with weak lensing in the 1990s, no surveys for new clusters were conducted, partly because of the small fields of view afforded by cameras on large telescopes until later in the decade, and partly because techniques needed to be proven on known clusters first. The first serendipitous detections of mass concentrations came when unexpected peaks appeared some distance from the target in mass reconstructions of known clusters.

In the first reported detection, a mass concentration was found projected near Abell 1942 (7′ from the center), and confirmed by a mass map constructed with data from a different camera and at a different wavelength [42]. There is no obvious concentration of galaxies associated with this mass, although the area does contain a poor group of galaxies and some weak X-ray emission. Because the redshift of the mass is unknown, its mass and M/L are also unknown. However, with an upper limit on the light in the area, the lower limit on M/L can be computed as a function of redshift. There are redshifts for which the object could have a reasonable M/L, around 400 [49]. This object is therefore not necessarily more dark than some X-ray selected clusters, which have M/L up to 600 or more [45,46] (see Sect. 2.2.3). If it is at the redshift of Abell 1942 ($z = 0.22$), its M/L is at least 600, which is very dark but just on the edge of the X-ray selected range, and perhaps explainable in a merger scenario with Abell 1942 proper.

In a second detection, Hubble Space Telescope (HST) imaging revealed an extra mass concentration one arcminute from the center of CL1604+4304 [128]. It is also seen in a second pointing shifted by $20''$, so it is likely to be real, but the interpretation is not clear. It seems likely to be substructure in the cluster rather than an independent structure, but the necessary followup is lacking. In a third case, serendipitous mass concentrations were found in a survey of known clusters [139]. Some of these corresponded to galaxy groups, which followup spectroscopy showed to be real and not associated with the target clusters, although in the same general redshift range (the range to which the lensing survey was of course most sensitive).

Although these hints were exciting, large "blank" fields (*i.e.* fields not selected to contain a known cluster) are more appropriate for finding unambiguously new clusters, and the first truly convincing shear-selected cluster was indeed discovered in such a field [138]. This object is clearly a cluster of galaxies (Fig. 2.13), with a solid spectroscopic redshift (0.28), velocity dispersion (615 km s^{-1}), and lens redshift coinciding with the spectroscopic value (see Sect. 2.2.6). The M/L is at the high end of, but definitely within, the range found for optically and X-ray selected clusters. Although clearly seen at visible wavelengths, no X-ray emission is detected at this position. A second shear-selected cluster has recently been assigned a spectroscopic redshift [137]. At $z = 0.68$, this cluster begins to fulfill the promise of lensing in terms of avoiding the r^{-2} falloff of methods which depend on emitted light.

Finally, the most recent candidate makes perhaps the strongest case yet for a dark cluster [94]. A tangential alignment was found around a point in a randomly selected $50''$ STIS field, significant enough that the data allow only a 0.3% chance of this occuring randomly. There are indications of strong lensing as well. A nearby group of galaxies could provide enough mass to explain this only if its M/L is two orders of magnitude higher than expected. As with the first two cases cited above, more followup, including a lens redshift, is desperately needed to make sense of this candidate.

Thus far, the serendipitous shear detections present no clear pattern, apart from the feeling that these are not typical optically or X-ray selected clusters. There is no proof of truly dark clusters, something which only lensing could detect. Perhaps this is only for lack of followup. Yet, the detection of a truly dark cluster would reveal surprisingly little about the nature of dark matter. Rather, it would indicate that either baryons did not fall into the potential well

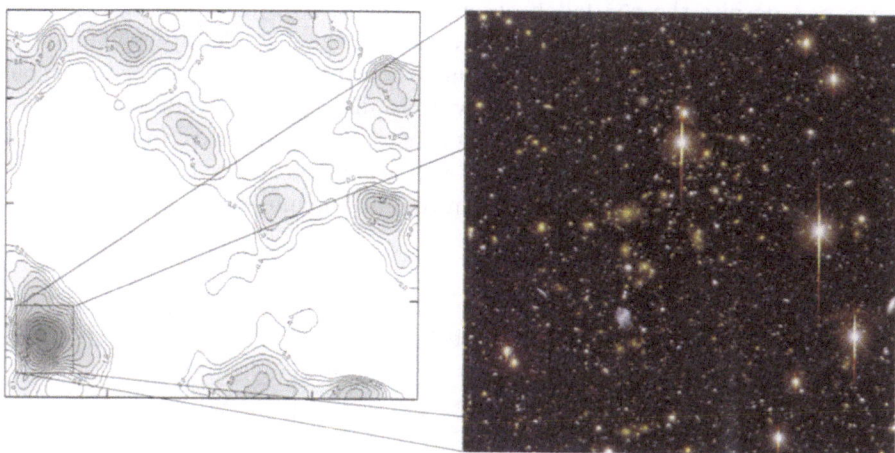

Fig. 2.13. A shear-selected cluster of galaxies. At left is a κ map of a $40'$ field not selected to contain a previously known cluster; black indicates higher density. The mass concentration at lower left corresponds to a cluster of galaxies (inset), spectroscopically confirmed at a redshift of 0.28 and a velocity dispersion of 615 km s^{-1} (see Color Plate).

created by the dark matter halo, or that star formation failed there. These are intriguing scenarios, but they raise questions about baryons rather than answer questions about dark matter.

Meanwhile, there are several surveys of tens of square degrees currently underway [127,37], which will yield samples of dozens of shear-selected clusters, rather than a serendipitous few, and perhaps yield a better idea of typical and extreme shear-selected clusters. Much work remains in terms of settling on estimators which maximize detection of real mass concentrations while minimizing false positives. For example, do we simply look for peaks in convergence maps (or maps of some other quantity such as potential or aperture mass), or do we apply a matched filter, which implies that we know what we are looking for ? While such work has been done theoretically and computationally [104], we must get our hands dirty with real samples before we can have much confidence in the scattered examples published as of today.

The advantages of shear selection in avoiding baryon and emitted-light bias are obvious, but no single cluster-finding technique will be completely unbiased. SZE selection [61] is an exciting new method which is also independent of emitted light. This is especially important in going to high redshift because of the r^{-2} falloff of emitted light. In this respect, SZE has the advantage because its background source is at a very high redshift (the cosmic microwave background at $z \sim 1100$), and because lensing is most efficient at detecting clusters at much lower redshift than the sources. However, lensing and SZE methods are so new that samples are not yet available. X-ray and optical selection are more established, and X-ray surveys have recently made great strides in detecting high-redshift clusters [20], indicating that it can compete with other methods at $z > 1$ despite the r^{-2} falloff of emitted light. X-ray emission has the additional advantage of depending on the square of the local density, making it less vulnerable to projection effects (although the density-squared dependence could be viewed as a disadvantage when trying to determine the total cluster mass). Table 2.1 summarizes the properties of these selection methods. In the end, comparison of differently selected samples will always be necessary, and much work remains to be done before we can claim that all the important biases are known.

Table 2.1. Comparison of cluster selection methods.

Selection method	Projection effects?	Emitted light?	Baryon dependent?	Samples available now?
Optical	yes	yes	yes	yes
X-ray	no	yes	yes	yes
Lensing	yes	no	no	almost
SZE	yes	no	yes	almost

2.2.6 Tomography with Clusters

Judging from the examples of the previous section, followup and identification of shear-selected clusters will be more difficult than finding them. The most basic parameter, redshift, is unknown in several cases. Without a redshift, even the lens mass and M/L must remain unknown, leaving little solid information. A spectroscopic redshift is impossible in the case of the Abell 1942 field with no obvious lens-associated galaxies, and difficult in the CL1604+4304 field, with CL1604+4304 itself projected so nearby. Thus, there is a great need for a method of determining the lens redshift from the lensing information alone.

If sources can be differentiated by redshift, the redshift of a lens will be revealed by the way that shear increases with source redshift (Fig. 2.3). Photometric redshifts are required for the sources, but this is straightforward if the deep imaging required for the shear measurement is extended to multiple filters. Two-filter imaging is routinely done anyway, to filter the sources based on color. Four filters is sufficient to provide photometric redshifts accurate to ~ 0.1 on each source, which is accurate enough given the large amount of shape noise on each galaxy and the breadth of the lensing kernel. This method has been demonstrated on one cluster [138] (Fig. 2.14). The most likely lens redshift is within 0.03 of the spectroscopic redshift ($z = 0.28$), but the formal error estimate is ~ 0.1.

Obviously, lens redshifts cannot compete with cluster spectroscopic, or even photometric, redshifts with this level of precision. Some improvement is to be expected as photometric redshifts improve. For example, the filter set used was not designed to be optimal for photometric redshifts, but future large surveys will be paying close attention to this issue. Also, the work cited neglected to use sources which were undetected in one or more filters, but photometric redshifts

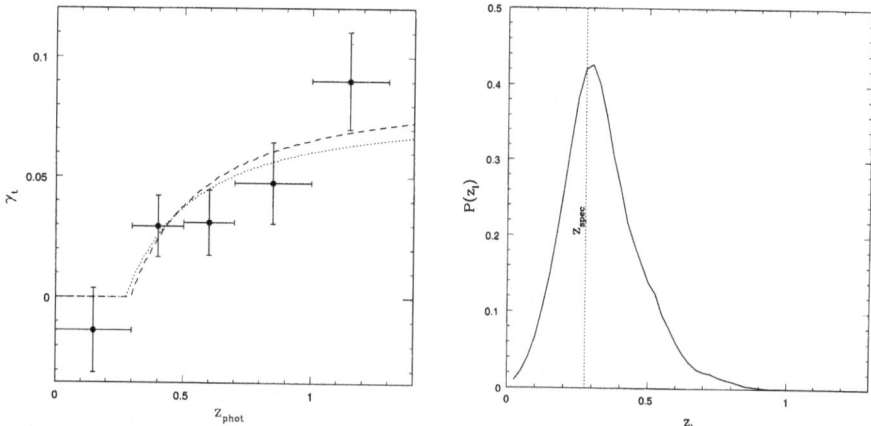

Fig. 2.14. *Left:* tangential shear as a function of source (photometric) redshift. The dotted line is the best fit for a lens at the cluster spectroscopic redshift of 0.28, while the dashed line is the best fit with the lens redshift derived from the lensing data alone. ($z = 0.30$). *Right:* The lens redshift probability distribution derived from the data at left. The method is promising: The most likely lens redshift is within 0.03 of the cluster spectroscopic redshift, but the width of the distribution is ~ 0.1, indicating the need for more precise data. From [138].

are not impossible to assign to such sources, and their inclusion could improve the statistics. A rigorous treatment would take account of each source's photometric redshift error estimate, and so on. Work is needed on optimal lens redshift algorithms.

Still, lens redshifts can be useful even at this level of accuracy. The most basic use is to confirm the unstated assumption in all cluster weak lensing work to date—that the cluster *is* the lens, not merely in the same line of sight. While no doubt valid, confirmation of such basic assumptions is always welcome. Second, in cases where dark mass concentrations are found without any associated galaxies, a lens redshift is the only way to constrain the basic parameters of the mass concentration. Indeed, if there is any skepticism about such claims, it would be conclusively dispelled by demonstrating that the observed shear increases with source redshift in the predicted way. Third, large weak lensing surveys may find enough shear-selected clusters to make complete spectroscopic followup burdensome. In that case, rough lens redshifts may be good enough for examining statistics of many clusters, or at least for identifying the more interesting candidates for followup. Therefore, this type of tomography will probably be a routine feature of future shear-selected surveys.

Finally, note how the spread in source redshifts would have caused more uncertainty in shear had source redshifts not been known in Fig. 2.14. One way of improving cluster shear measurements, which until now have used at most a color cut to avoid contamination of the sources by cluster members, will be the use of photometric redshifts to weight sources. If Fig. 2.14 is any guide, this might make an improvement of up to a factor of two.

2.3 Large-Scale Structure

Clusters are not the largest structures in the universe. Although it had long been known that clusters themselves tend to cluster, it was only in the 1980's that redshift surveys began to reveal apparently coherent structures—filaments and voids—on very large scales, up to $\sim$ 50 Mpc. Current redshift surveys are producing impressive views of this foamy galaxy distribution out to $\sim$ 600 Mpc, or $z \sim 0.2$ [100]. But what about the mass distribution ?

Simulations of cold dark matter show similar structures in mass (Fig. 2.15). Furthermore, they show how the evolution of large-scale structure depends on cosmological parameters and on the nature of dark matter. Good measurements of large-scale structure evolution should therefore be able to constrain cosmological parameters and the nature of dark matter through comparison with simulations. Weak lensing is a good candidate for such comparisons, because like the simulations it deals with mass, not galaxies; and because it can easily reach back to $z \sim 1 - 2$, providing a long baseline in cosmic time.

Fig. 2.15 illustrates just how many voids and filaments are expected to lie between us and a source at $z \sim 1$. Because of projection effects, weak lensing will never produce stunningly detailed three-dimensional mass maps to allow comparisons with such simulations. But weak lensing by large-scale structure

Fig. 2.15. Simulation from the Virgo collaboration showing the evolution of large-scale structure in a $7°$ slice of a Λ-dominated universe, with black indicating highest density. The dotted lines indicate $z = 1$ and $z = 2$. Adapted from [133].

does leave a statistical signature. This "cosmic shear" is a strong function of cosmological parameters, in particular, Ω_m and σ_8, the rms density variation on 8 Mpc scales, and thus is potentially a very useful cosmological tool. However, cosmic shear leaves a much weaker signal than do clusters, making detection more difficult and systematics more dangerous. The first detections of cosmic shear came in 2000, a decade after weak lensing by clusters was detected. For that reason, cosmic shear is just beginning to take its place in the cosmology toolbox.

2.3.1   Cosmic Shear Estimators

Cosmic shear, unlike the shear induced by clusters and groups, has no center, and that has led to the formulation of a variety of statistics different from those used to analyze clusters. We summarize them here to provide the basis for interpreting the results presented here and in the literature. Each of the following statistics has advantages and disadvantages, and current practice is to report results in terms of several different estimators to verify robustness.

A few comments apply to all the estimators mentioned below. Current wide-field cameras have fields of view of $\sim 0.5°$, and all results to date have been reported on these or smaller scales. But a look at Fig. 2.15, with its opening angle of 7°, shows that such small fields will give different results depending on where they happen to lie. Because of this *cosmic variance*, one such field cannot really constrain the cosmology. A sample of randomly chosen fields is required, with the field-to-field scatter in results giving some idea of the cosmic variance. Observed variance could also be due to problems with the instrument or telescope, so this is really an upper limit to the cosmic variance, but it is still a very useful number. Some groups are currently doing much larger fields by stitching together multiple pointings, sometimes with sparse sampling, but multiple, widely separated fields are still required to insure that cosmic variance has been beaten.

On small scales, the dominant statistical noise source is simply shape noise, but systematics are also larger here. Small-scale PSF variations cannot be mapped because the density of stars is too low; intrinsic alignments play a larger role on small scales; and comparison to theory (not necessarily simulations) is hampered by the difficulty of modeling the nonlinear collapse of dense regions. Results on scales $< 1'$ may say more about nonlinear collapse and possibly intrinsic alignments than about the cosmology.

Mean Shear. The mean shear in a field (simply averaging all sources) will in general be nonzero in the presence of lensing. However, it will tend to zero for a field of any significant size, so this statistic is of limited use. We mention it for completeness, as some early work with small fields of view used this statistic. But mean shear in a small field of view is difficult to interpret, as it could result from a single structure projected near the line of sight.

Shear Variance. The next logical step is to compute the variance, among a group of boxes of angular size ϕ, of the mean shear in each box. Because variance is a positive definite quantity, noise rectification must be subtracted off. The importance of the noise rectification term depends on the number of sources per box. Roughly speaking, for $\phi < 1'$, the noise rectification term is larger than the lensing signal itself, but for larger ϕ the lensing contribution dominates and at several arcminutes the noise correction becomes quite small. Thus, measurements of shear variance at $\phi < 1'$ should be treated with caution (in addition to the cautions cited above for small scales). Usually the results are presented as a function of ϕ, but the values for different ϕ have been computed from the same data. Hence the errors on the different scales are not independent, and results are sometimes less significant than appears at first glance. The true significance of such results can be explored with bootstrap resampling. A rule of thumb suggested by bootstrap tests is that measurements on widely differing scales (factors of 10) are largely independent of each other even when computed from the same data.

Ellipticity Correlations. The observed ellipticities of lensed sources are correlated, so it is natural to construct a correlation function which measures this effect as a function of angular separation between sources. A simple correlation of the ellipticity components e_+ and $e_\times$ would have little physical meaning, though, as they depend on the orientation of the detector axes. If the components of a pair of galaxies are instead defined with respect to an imaginary line joining their centers, their correlation does have a physical significance [92]. In fact, three useful functions can be defined:

$$\xi_1(\theta) \equiv \langle e_+^i e_+^j \rangle$$
$$\xi_2(\theta) \equiv \langle e_\times^i e_\times^j \rangle \qquad (2.8)$$
$$\xi_3(\theta) \equiv \langle e_+^i e_\times^j \rangle$$

where superscripts label the sources and brackets denote averaging over all pairs of galaxies $i \neq j$ with angular separation θ.

Like shear variance, lensing induces $\xi_1 > 0$ for all θ, but decreasing with θ. Unlike shear variance, the computation of ξ_1 does *not* result in a positive definite quantity, so spurious results may be easier to identify. The behavior of ξ_2 in the presence of lensing by large-scale structure is more interesting: at $\theta = 0$, ξ_2 and ξ_1 are equal, but ξ_2 drops more rapidly and goes negative at some θ which depends on the cosmology ($\sim 0.5 - 1°$). ξ_3 is a control statistic. Unaffected by lensing, it should vanish in the absence of systematic errors or intrinsic alignments (this is equivalent to rotating one of each pair of galaxies by $45°$). Taken together, these properties provide a signature with several lines of defense against systematic error.

Like shear variance, values for different θ are computed from the same data, so the same warnings about nonindependent angular bins apply. Unlike shear variance, though, there are two independent quantities (ξ_1 and ξ_2) at each θ, which can be checked against each other. For example, unless $\xi_1(0) = \xi_2(0)$ and

$\xi_1(\theta) > \xi_2(\theta)$ for $\theta > 0$, the results are suspect. If these checks make sense, ξ_1 and ξ_2 can be combined into a single higher signal-to-noise measurement. In fact, shear variance can be understood as $\xi_1(0) + \xi_2(0)$, convolved with a square window function of width ϕ.

Aperture Mass. The aperture mass statistic M_{ap} was designed to address the nonlocality of shear. It is a generalization of the ζ statistic already mentioned in the context of clusters [45,73]. As in the ζ statistic, tangential shear is computed in a circular aperture, but here it is weighted with a compensated filter function; the weight is positive at the center of the aperture and negative at the edges, for a total weight of zero. This has the remarkable property of making adjacent apertures nearly independent, whereas the shear in adjacent apertures is highly correlated. M_{ap} was first suggested as a way of looking for clusters in wide-field images [112] , and later its variance $\langle M_{ap}^2 \rangle$ was proposed as a measure for cosmic shear [113] (note that because the total weight vanishes, so does the expectation value: $\langle M_{ap} \rangle = 0$). Although aperture mass tends to be noisier than the other estimators, its compensating virtue is that measurements on different scales are almost completely independent.

Other Estimators. All of the above estimators are at most two-point statistics. Higher-order statistics have been proposed. For example, values of the projected mass field (κ or M_{ap}) should have a skewness due to many somewhat underdense regions (voids) and a few extreme overdense regions. This skewness depends on Ω_m and the matter power spectrum in a different way than does the variance, leading to suggestions that together they could constrain both quantities [14]. This non-Gaussianity may also be revealed through morphological analysis of convergence fields [110,121,52]. All this remains largely theoretical, as these high-order statistics are in practice noisier than the two-point estimators which are providing the first detections of cosmic shear. There is a recent claim of detection of this non-Gaussian signature [13], but an accurate measurement of skewness requires that rare massive halos be present in the sample [32], hence a very large area is required. The best way to the power spectrum itself may be through maximum likelihood fitting to the shear data [66,31].

2.3.2 Observational Status

Although the idea of weak lensing by large-scale structure was first suggested in the 1960's [80,53], the effect escaped detection for over three decades. The first attempts at detection gave null results [79,130], which is not surprising given the subtleness of the effect ($\sim 1\%$ shear) and the lack of sensitivity and nonlinearity of photographic plates. The first analysis of CCD data, albeit with the narrow field afforded by CCDs in 1994, also yielded only upper limits [97]. With the advent of large-format CCD mosaics, detection was inevitable, and four groups [140,131,6,75] announced detections in the span of one month in 2000.

Their results are summarized in terms of shear variance in Fig. 2.16. The groups used four different cameras on three different telescopes, with different

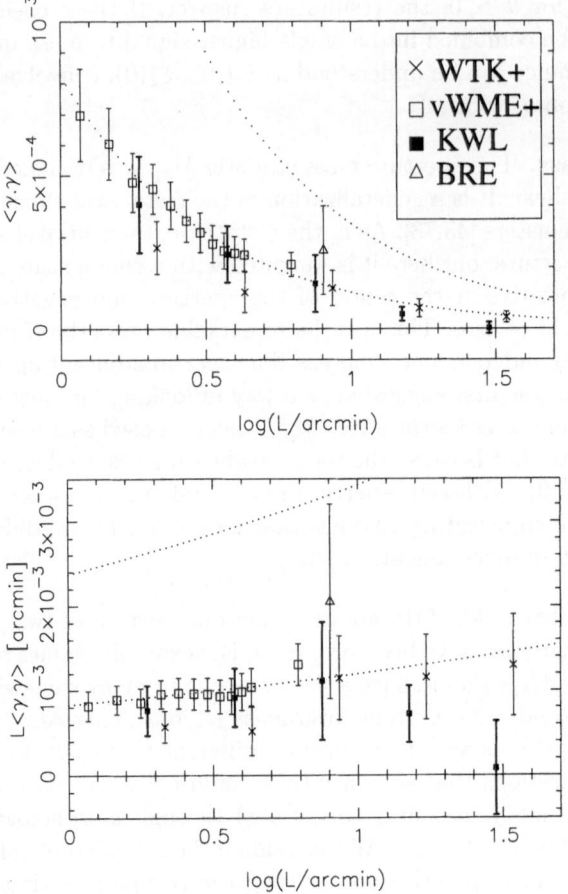

Fig. 2.16. First detections of cosmic shear, in terms of shear variance versus angular scale. The results of four different groups using three different telescopes and four different cameras are shown, with good agreement. Note that different angular bins from the same experiment are not independent. The dotted lines are for two different source redshift distributions (lower, $\langle z \rangle = 1$; upper, $\langle z \rangle = 2$) in a ΛCDM universe. Adapted from [75].

observed bandpasses and data reduction procedures and analysis techniques, yet the results were in good agreement. This has been taken as proof that instrumental effects and systematic errors have been vanquished, but in fact, the results should *not* agree if the data and source selection resulted in different mean source redshifts. The fact that the areal density of sources used was similar for all four groups suggests that the source redshifts were similar despite the different approaches. But the possibility remains that different source redshifts are hiding some disagreement.

Nevertheless, all results point to a low-Ω_m universe. Figure 2.16 shows the good fit to ΛCDM. It is difficult to constrain Λ with these measurements, but shear variance should scale roughly with Ω_m, so it is clear that $\Omega_m = 1$, for example, is ruled out. While this was no surprise, it signaled the emergence of cosmic shear as a new way to constrain Ω_m, completely independent of traditional methods (supernovae, CMB, age of the oldest stars in conjunction with the Hubble constant, etc.).

Since then, there have been further detections both in ground-based [87,58] and space-based data [105]. The state of the art is a many-sigma detection (whatever estimator is chosen) over 6.5 deg^2 leading to quantitative constraints in the Ω_m, σ_8 plane [132]. There is a significant degeneracy between Ω_m and σ_8; the first generation of cosmic shear papers simply assumed a value of σ_8 consistent with the local abundance of clusters. At the same time, efforts to improve the signal-to-noise of cosmic shear measurements by decomposition of cosmic shear into E and B modes are underway and appear to have met with success [103].

Currently, several large (tens of square degrees) surveys are under way with the goal of very high signal-to-noise analyses of cosmic shear [127,37,58]. At the same time, the question of how accurate the measurements can ultimately get is being explored [5,43,82]. However, a word of caution is in order. Such analyses tend to ignore the fact that the source redshift distribution is not well known. The putative accuracy of current and near-future cosmic shear measurements thus tends to be far too optimistic. For the moment, the most accurate measurements in an absolute sense will be those which are no deeper than current redshift surveys. Within a few years, though, this will probably change as photometric redshifts are used to estimate the source redshift distribution accurately enough. Photometric redshifts, in fact, will enable probing the redshift evolution of cosmic shear; division of sources into just two or three redshift bins can greatly improve the measurements of cosmological parameters, specifically Ω_Λ by a factor of ~ 7 [62].

2.4   Future Prospects

2.4.1   New Applications

It is impossible to predict what new applications weak lensing might find, but it is worth discussing one example of an interesting new direction: constraints on theories of gravity. It is unlikely that weak lensing will serendipitously reveal some new feature of gravity, because the lenses through which we look are not well calibrated. But given an alternative theory of gravity, we can ask if weak lensing observations are consistent with other observations.

Modified gravity is an attempt to explain differences between light distributions and inferred mass distributions without invoking dark matter. It is possible to modify Newtonian gravity to account for some of the observed differences such as flat rotation curves in galaxies, but a general correlation between mass and light remains. If lensing were to find severe discrepancies between mass and light, such as a dark cluster or clear misalignment of cluster mass and light axes, this would represent a serious blow to modified gravity [115]. There are some promising dark cluster candidates (Sect. 2.2.5), and Abell 901b presents mass and light axes which apparently differ [50], but there are no bulletproof examples. Weak lensing surveys of significant areas are only now underway, so it will take some time before dark clusters can be ruled in or out with much confidence. Note that dark clusters are not *expected* in dark matter scenarios; mass concentrations should accumulate enough baryons to become visible, if only in X-rays. Thus an absence of dark clusters would not favor modified gravity over dark matter, but their presence of would disprove modified gravity as currently envisioned.

Recently, the first quantitative predictions of weak lensing in modified gravity scenarios were published. Modified gravity, by increasing the strength of gravity on large scales, would greatly enhance cosmic shear, inconsistent with measurements. Thus, at the large scales probed by cosmic shear, the r^{-2} force law cannot be modified—if gravity does depart from r^{-2}, it is only on scales from 10 h^{-1} kpc to 1 h^{-1} Mpc [134]. Weak (and strong) lensing can also address modified gravity by constraining halo flattening [96]. Weak lensing by large-scale structure can also provide a test of higher-dimensional gravity [129].

2.4.2   New Instruments

Today's surveys of tens of square degrees will take years to find perhaps dozens of shear-selected clusters and put some constraints on w, the dark energy equation of state, as well as Ω_m [55]. Tight constraints on both would require a very deep survey of 1000 deg^2 [67], taking decades with current telescopes and instruments. The latest generation of 8-m class telescopes does not really help, as their fields of view are small, typically $\sim 10'$ or less (an exception is the Subaru telescope which has a $\sim 24'$ field of view with its SuPrime camera, a likely source of weak-lensing results in the near future). A new generation of wide-field telescopes specifically designed for surveys will dramatically accelerate our ability to do cosmology with large lensing surveys. These surveys will cover an area comparable to that of SDSS, but much more deeply. The first of these to be funded is VISTA, a 4-m telescope with a 1° field of view currently in the design stage, but apparently it will be infrared-only, limiting its usefulness for weak lensing. LSST, an 8-m class telescope with a 3° field of view [126] and concentrating on visible wavelengths, may also be built within a decade.

Figure 2.17 shows the potential of a 1000 deg^2 survey which LSST could easily accomplish. Of course, predictions such as these depend on the extrapolation of $\sqrt{n}$ statistics to extremely large areas, so it is wise to ask what systematic effects might provide a higher noise floor. Early work on cosmology constraints from cluster counts assumed NFW profiles for all clusters [81]. It was then realized that the profile makes a big difference, so that cluster counts may tell us more about dark-matter profiles than about cosmology [9]. However, new estimators have been proposed to circumvent this problem [55]. Careful attention must also be paid to the issue of completeness versus false positives in cluster-detection surveys [135]. Still, by the time LSST starts operation, these issues may be worked out, and it may be well to survey all the sky visible from the site. Such

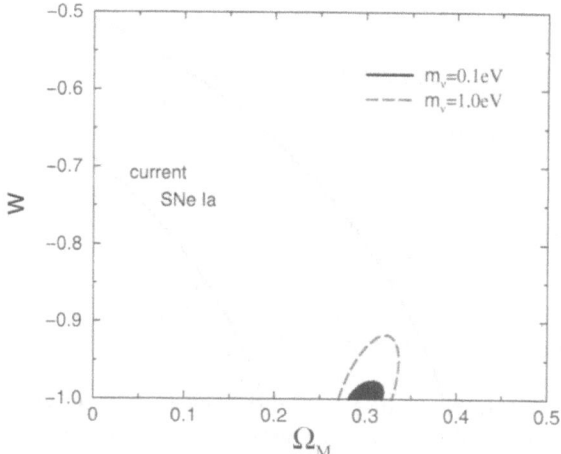

Fig. 2.17. 68% confidence limit constraints on Ω_M and w for two values of m_ν, for a weak lensing survey of 1000 deg^2 down to $R = 27$, with photometric redshifts providing the source redshift distribution. Current 1-σ constraints from type Ia supernovae are shown for comparison. From [67].

a survey would also provide a shear power spectrum comparable in accuracy to the CMB power spectra of today.

Another probe of cosmology which may become feasible with such massive surveys involves the angular power spectrum of clusters. The linear part of this power spectrum is essentially a standard ruler calibrated by the CMB, so that a power spectrum of clusters at a particular redshift yields the angular diameter distance to that redshift. A very large survey (~ 4000 deg^2) could determine this as a function of redshift, which of course would yield an absolute calibration of the distance scale and the Hubble constant [33].

2.4.3   New Algorithms

The combination of lensing data with other types of data has been an active theoretical area recently, and some of these algorithms will soon prove themselves observationally. Lensing plus SZE measurements of clusters will reveal the baryon fraction in that environment, perhaps leading to a new estimate of Ω_m from baryon scaling arguments—or perhaps leading to new aspects of cluster formation. Combinations of lensing and SZE plus X-ray data will help deproject cluster mass and gas distributions [142].

Cross-correlation of lensing by large-scale structure with the CMB will reveal parameters largely hidden from traditional CMB analyses, such as dark energy, the end of the dark ages, and the gravitational wave amplitude [64,65]. Lensing of the CMB itself may be detectable by the Planck satellite and constrain the amplitude of mass fluctuations between us and $z \sim 1000$ [120], but this may have to wait for even higher-sensitivity CMB probes [63].

References

1. H. Abdelsalam, P. Saha, L. Williams: AJ **116**, 1541 (1998)
2. G. Abell, H. Corwin, R. Olowin: ApJS **70**, 1 (1989)
3. S. Allen: MNRAS **296**, 392 (1998)
4. S. W. Allen, R. W. Schmidt, A. C. Fabian: MNRAS **328**, L37 (2001)
5. D. Bacon, A. Refregier, D. Clowe, R. Ellis: MNRAS **325**, 1065 (2001)
6. D. Bacon, A. Refregier, R. Ellis: MNRAS **318**, 625 (2000)
7. N. Bahcall, R. Cen, R. Davé, J. Ostriker, Q. Yu: ApJ **541**, 1 (2000)
8. C. Balland, A. Blanchard: ApJ **487**, 33 (1997)
9. M. Bartelmann, L. King, P. Schneider: A&A **378**, 361 (2001)
10. M. Bartelmann: A&A **303**, 643 (1995)
11. M. Bartelmann, R. Narayan: ApJ **451**, 60 (1995)
12. M. Bartelmann, R. Narayan, S. Seitz, P. Schneider: ApJ **473**, 610 (1996)
13. F. Bernardeau, Y. Mellier, L. van Waerbeke: ApJL, submitted, astro-ph/0201032 (2002)
14. F. Bernardeau, L. van Waerbeke, Y. Mellier: A&A **322**, 1 (1997)
15. G. Bernstein, M. Jarvis: AJ **123**, 583 (2002)
16. A. Biviano: to appear in "Tracing Cosmic Evolution with Galaxy Clusters", ASP Conference Series; astro-ph/0110053 (2001)
17. J. Blakeslee: astro-ph/0108253 (2001)
18. R. Blandford, A. Saust, T. Brainerd, J. Villumsen: MNRAS **251**, 600 (1991)
19. P. Bode, N. Bahcall, E. Ford, J. Ostriker: ApJ **551**, 15 (2001)
20. S. Borgani, L. Guzzo: Nature **409**, 39 (2001)
21. T. Brainerd, C. Wright, D. Goldberg, J. Villumsen: ApJ **524**, 9 (1999)
22. T. Broadhurst, A. Taylor, J. Peacock: ApJ **438**, 49 (1995)
23. M. Brown, A. Taylor, N. Hambly, S. Dye: MNRAS, submitted, astro-ph/0009499 (2000)
24. S. Burles, K. Nollett, M. Turner: ApJL **552**, L1 (2001)
25. P. Catelan, M. Kamionkowski, R. Blandford: MNRAS, **320L**, 7 (2001)
26. R. Cen: ApJ **485**, 39 (1997)
27. T.-C. Chang, A. Refregier: ApJ, **570**, 447 (2002)
28. D. Clowe, P. Schneider: A&A **379**, 384 (2001)
29. J. Cohen, D. Hogg, R. Blandford, L. Cowie, E. Hu, A. Songaila, P. Shopbell, K. Richberg: ApJ **538**, 29 (2000)
30. A. Connolly, I. Csabai, A. Szalay, D. Koo, R. Kron, J. Munn: AJ **110**, 2655 (1995)
31. A. Cooray, W. Hu: ApJ **554**, 56 (2001)
32. A. Cooray, W. Hu: ApJ **548**, 7 (2001)
33. A. Cooray, W. Hu, D. Huterer, M. Joffre: ApJL **557**, 7 (2001)
34. R. Crittenden, P. Natarajan, U. Pen, T. Theuns: ApJ **559**, 552 (2001)
35. R. Crittenden, P. Natarajan, U. Pen, T. Theuns: ApJ **568**, 20 (2002)
36. R. Croft, C. Metzler: ApJ **545**, 561 (2000)
37. http://terapix.iap.fr/Descart
38. M. Donahue, G. Voit, I. Gioia, G. Luppino, J. Hughes, J. Stocke: ApJ **502**, 550 (1998)
39. O. Dore, F. Bouchet, Y. Mellier, R. Teyssierm: A&A **375**, 14 (2001)
40. S. Dye, A.N. Taylor, T.R. Greve, O.E. Rognvaldsson, E. van Kampen, P. Jakobsson, V.S. Sigmundsson, E.H. Gudmundsson, J. Hjorth: A&A **386**, 12 (2002)
41. H. Ebeling, L.R. Jones, B.W. Fairley, E. Perlman, C. Scharf, D. Horner: ApJL **548**, 23 (2001)

42. T. Erben, L. van Waerbeke, Y. Mellier, P. Schneider, J.-C. Cuillandre, F.J. Castander, M. Dantel-Fort: A&A **355**, 23 (2000)
43. T. Erben, L. van Waerbeke, E. Bertin, Y. Mellier, P. Schneider: A&A **366**, 717 (2001)
44. E. Falco, M. Gorenstein, I. Shapiro: ApJ **289**, L1 (1985)
45. Fahlman, G., Kaiser, N., Squires, G., & Woods, D.: ApJ **437**, 56 (1994)
46. P. Fischer: AJ **117**, 2024 (1999)
47. P. Fischer, J. A. Tyson: AJ **114**, 14 (1997)
48. D. Goldberg, M. Natarajan: ApJ, submitted, astro-ph/0107187 (2001)
49. M. Gray, R. Ellis, J. Lewis, R. McMahon, A. Firth: MNRAS **325**, 111 (2001)
50. M. Gray, A. Taylor, K. Meisenheimer, S. Dye, C. Wolf, E. Thommes: ApJ **568**, 141 (2002)
51. L. Grego, J. Carlstrom, E. Reese, G. Holder, W. Holzapfel, M. Joy, J. Mohr, S. Patel: ApJ **552**, 2 (2001)
52. A. Guimaraes: MNRAS, submitted, astro-ph/0202507 (2002)
53. J. Gunn: ApJ **147**, 61 (1967)
54. A. Heavens, A. Refregier, C. Heymans: MNRAS **319**, 649 (2000)
55. J. Hennawi, V. Narayanan, D. Spergel, I. Dell'Antonio, V. Margoniner, J. A. Tyson, D. Wittman: BAAS **199**, 1608 (2001)
56. H. Hoekstra: A&A **370**, 743 (2001)
57. H. Hoekstra, M. Franx, K. Kuijken, R.G. Carlberg, H.K.C. Yee, H. Lin, S.L. Morris, P.B. Hall, D.R. Patton, M. Sawicki, G.D. Wirth: ApJL **548**, L5 (2001)
58. H. Hoekstra, H. Yee, M. Gladders: to appear in the proceedings of the STScI 2001 spring symposium "Dark Universe", astro-ph/0106388 (2001)
59. D. Hogg: astro-ph/9905116 (1999)
60. D. Hogg, *et al.*: AJ **115**, 1418 (1998)
61. G. Holder, J. Mohr, J. Carlstrom, A. Evrard, E. Leitch: ApJ **544**, 629 (2000)
62. W. Hu: ApJL **522**, L21 (1999)
63. W. Hu: PRD **62**, 3007 (2000)
64. W. Hu: ApJL **557**, L79 (2001)
65. W. Hu: Phys. Rev. D, submitted, astro-ph/0108090 (2001)
66. W. Hu, M. White: ApJ **554**, 67 (2001)
67. D. Huterer: ApJ, submitted, astro-ph/0106399 (2001)
68. N. Kaiser: ApJ **388**, 272 (1992)
69. N. Kaiser: ApJ **498**, 26 (1998)
70. N. Kaiser: in *Gravitational Lensing: Recent Progress and Future Goals*, ASP Conference Proceedings, Vol. 237. Edited by Tereasa G. Brainerd and Christopher S. Kochanek. San Francisco: Astronomical Society of the Pacific, ISBN: 1-58381-074-9, p.269 (2001)
71. N. Kaiser: ApJ **537**, 555 (2000)
72. N. Kaiser, G. Squires: ApJ **404**, 441 (1993)
73. N. Kaiser, G. Squires, T. Broadhurst: ApJ **449**, 460 (1995)
74. N. Kaiser, G. Wilson, G. Luppino, L. Kofman, I. Gioia, M. Metzger, H. Dahle: ApJ, submitted, astro-ph/9809268 (1998)
75. N. Kaiser, G. Wilson, G. Luppino: ApJL, submitted, astro-ph/0003338 (2000)
76. C. Keeton: astro-ph/0102341 (2001)
77. L. King, P. Schneider: A&A **369**, 1 (2001)
78. A. Knebe, R. Islam, J. Silk: MNRAS **326**, 109 (2001)
79. J. Kristian: ApJ **147**, 864 (1967)
80. J. Kristian, R. Sachs: ApJ **143**, 379 (1966)

81. G. Kruse, P. Schneider: MNRAS **302**, 821 (1999)
82. K. Kuijken: astro-ph/0007368 (2000)
83. J. Lee, U. Pen: ApJ **555**, 106 (2000)
84. J. Lee, U. Pen: ApJL **567**, L111 (2000)
85. R. Lynds, V. Petrosian: BAAS **18**, 1014 (1986)
86. J. Mackey, M. White, M. Kamionkowski: MNRAS, **332**, 788 (2002)
87. R. Maoli, L. van Waerbeke, Y. Mellier, P. Schneider, B. Jain, F. Bernardeau, T. Erben, B. Fort: A&A **368**, 766 (2001)
88. C. Mayen, G. Soucail: A&A **361**, 415 (2000)
89. Y. Mellier: ARAA **37**, 127 (1999)
90. C. Metzler, M. White, C. Loken: ApJ **547**, 560 (2001)
91. J. Miralda-Escudé: ApJ **370**, 1 (1991)
92. J. Miralda-Escudé: ApJ **380**, 1 (1991)
93. J. Miralda-Escudé, A. Babul: ApJ **449**, 18 (1995)
94. J.-M. Miralles, T. Erben, H. Haemmerle, P. Schneider, R.A.E. Fosbury, W. Freudling, N. Pirzkal, B. Jain, S.D.M. White: A&A submitted, astro-ph/0202122 (2002)
95. R. Moessner, B. Jain: MNRAS **294**, 291 (1998)
96. D. J. Mortlock, E. L. Turner: MNRAS **327**, 552 (2001)
97. J. Mould, R. Blandford, J. Villumsen, T. Brainerd, I. Smail, T. Small, W. Kells: MNRAS **271**, 31 (1994)
98. J. Navarro, C. Frenk, S. White: MNRAS **275**, 720 (1995)
99. J. Navarro, C. Frenk, S. White: ApJ **490**, 493 (1997)
100. J. Peacock *et al.*: Nature **410**, 169 (2001)
101. P. J. E. Peebles: *Principles of Physical Cosmology*, Princeton University Press (1993)
102. U. Pen, Li, U. Seljak: ApJL **543**, L107 (2000)
103. U. Pen, L. van Waerbeke, Y. Mellier: ApJ **567**, 31 (2002)
104. K. Reblinsky, M. Bartelmann: A&A **345**, 1 (1999)
105. J. Rhodes, A. Refregier, E. Groth: ApJL **552**, 85 (2001)
106. J. Robinson, E. Gawiser, J. Silk: ApJ **532**, 1 (2000)
107. A. Refregier: MNRAS, submitted, astro-ph/0105178 (2001)
108. A. Refregier, D. Bacon: MNRAS, submitted, astro-ph/0105179 (2001)
109. W. Saslaw: *Gravitational Physics of Stellar and Galactic Systems*, Cambridge University Press (1987)
110. J. Sato, M. Takada, Y. P. Jing, T. Futamase: ApJL **551**, L5 (2001)
111. P. Schneider, S. Seitz: A&A **294**, 411 (1995)
112. P. Schneider: MNRAS **283**, 837 (1996)
113. P. Schneider, L. van Waerbeke, B. Jain, G. Kruse: MNRAS **296**, 873 (1998)
114. P. Schuecker, H. Bohringer, H. Reiprich, L. Feretti: A&A **378**, 408 (2001)
115. J. Sellwood, A. Kosowsky: to appear in *The Dynamics, Structure & History of Galaxies*, G. S. Da Costa & E. M. Sadler, eds, ASP Conference Series; astro-ph/0109555 (2002)
116. E. Sheldon, *et al.*: ApJ **554**, 881 (2001)
117. I. Smail, R. Ellis, A. Dressler, W. Couch, A. Oemler, R. Sharples, H. Butcher: ApJ **479**, 70 (1997)
118. G. Soucail, B. Fort, Y. Mellier, J. Picat: A& A **172**, L14 (1987)
119. G. Squires, N. Kaiser: ApJ **473**, 65 (1996)
120. M. Takada, T. Futamase: ApJ **546**, 620 (2001)
121. A. Taruya, M. Takada, T. Hamana, I. Kayo, T. Futamase: ApJ, submitted, astro-ph/0202090 (2002)

122. J. A. Tyson: AJ **96**, 1 (1988)
123. J. A. Tyson, G. Kochanski, I. Dell'Antonio: ApJ **498**, 107 (1998)
124. J. A. Tyson, P. Seitzer: ApJ **335**, 552 (1988)
125. J. A. Tyson, R. Wenk, F. Valdes: ApJL **349**, L1 (1990)
126. J. A. Tyson, D. Wittman, J. R. P. Angel: to appear in proceedings of the Dark Matter 2000 conference (Santa Monica, February 2000) to be published by Springer, astro-ph/0005381 (2000)
127. J. A. Tyson, D. Wittman, I. Dell'Antonio, A. Becker, V. Margoniner, DLS Team: BAAS **199**, 10113 (2001)
128. K. Umetsu, T. Futamase: ApJL **539**, L5 (2000)
129. J.-P. Uzan, F. Bernardeau: Phys. Rev. D 64, 3004 (2001)
130. F. Valdes, J. Jarvis, J. A. Tyson: ApJ **271**, 431 (1983)
131. L. van Waerbeke *et al.*: A&A **358**, 30 (2000)
132. L. van Waerbeke *et al.*: A&A **374**, 757 (2001)
133. http://www.mpa-garching.mpg.de/Virgo/virgoproject.html
134. M. White, C. S. Kochanek: ApJ **560**, 539 (2001)
135. M. White, L. van Waerbeke, J. Mackey: astro-ph/0111490 (2001)
136. G. Wilson, N. Kaiser, G. Luppino: ApJ **556**, 601 (2001)
137. D. Wittman, V. Margoniner, J. A. Tyson, J. Cohen, I. Dell'Antonio: ApJL in prep (2002)
138. D. Wittman, J. A. Tyson, V. Margoniner, J. Cohen, I. Dell'Antonio: ApJL **557**, L89 (2001)
139. D. Wittman, I. Dell'Antonio, J. A. Tyson, G. Bernstein, P. Fischer, D. Smith: in *Constructing the Universe with Clusters of Galaxies*, IAP 2000 meeting, Paris, France, July 2000, Florence Durret & Daniel Gerbal (Eds.) (2000)
140. D. Wittman, J. A. Tyson, D. Kirkman, I. Dell'Antonio, G. Bernstein: Nature **405**, 143 (2000)
141. D. York *et al.*: AJ **120**, 1579 (2000)
142. S. Zaroubi, G. Squires, G. de Gasperis, A. Evrard, Y. Hoffman, J. Silk: ApJ **561**, 600 (2001)

3 Gravitational Optics Studies of Dark Matter Halos

Tereasa G. Brainerd[1] and Roger D. Blandford[2]

[1] Boston University, Department of Astronomy, 725 Commonwealth Avenue, Boston, MA 02215 USA
[2] California Institute of Technology, Theoretical Astrophysics 130-33, Pasadena, CA, 91125, USA

Abstract. One of the best methods that we have for exploring the distribution of dark matter throughout the universe is to measure its gravitational influence on the observed shapes of background galaxies. On linear scales of $\sim 3 - 300$ kpc, when the dark matter distribution is dominated by individual galaxy halos, the investigation is best performed statistically by correlating the orientations of the source galaxies with the position angles of the foreground lens galaxies. Recent observational programs have shown that the large-scale mass to light ratio in red bandpasses is ~ 100 solar units, and is apparently independent of galaxy type. The prospects for improving the precision of these measurements and for extending them to larger length scales over the next five years are good.

3.1 Introduction

Gravitational optics, first considered in an astronomical context by Soldner [42] and Einstein [10], [11], and first definitively observed by Walsh et al. [48], has evolved from a fascinating optical curiosity to a rapidly maturing set of cosmological techniques. Its many possible applications include:

- measuring the size and shape of the universe
- quantifying the large-scale distribution of dark matter in the universe
- measuring the gravitational potential in clusters of galaxies
- investigating the dark matter content and extent of galaxy halos
- probing the interstellar media of intervening galaxies and the intergalactic medium
- observing intrinsically faint sources with high angular resolution and magnification

Each of these applications has been realized and in some of these examples, gravitational optics now provides the primary and most reliable tool available to the observational cosmologist.

In considering gravitational optics, two distinct cases can be distinguished: strong lensing, for which multiple images are produced, and weak lensing, for which only mild image distortion occurs. Whenever possible, instances of strong and weak lensing are combined in order to get the full picture. However, in this article we will focus on only one particular technique: weak lensing by galaxy halos. Although this is a relatively new approach, it has already advanced to the point where it has yielded unexpected results.

The primary scientific goal of this work is to map the shapes of the dark halos of distant galaxies statistically, which can be accomplished by measuring the shapes of background galaxies and seeking a correlation with the positions of foreground galaxies. In particular, we want to measure (and define) the masses of galaxies and the slopes of their density distributions. In addition, we would like to determine the amount of flattening of the dark potential and its relationship to the light distribution, as well as its variation with radius. Ultimately, of course, we would like to understand the association of all of these characteristics with galaxy morphology, environment and cosmological epoch.

In what follows, we shall describe the methodology of weak lensing and the manner in which galaxy halos are characterized so as to learn about their properties. We shall emphasize the current uncertainties associated with defining dark halos at large and small radius and suggest that these are areas where a large improvement will soon be possible. We shall also highlight recent research on the feasibility of measuring the shapes of halos and of distinguishing the shapes of galaxies in clusters from those in the field. Finally, we shall outline two approaches to measuring the redshift distribution of the faint and the ultrafaint sources seen on the deepest HST images.

For recent comprehensive reviews of weak lensing and its applications, we refer the reader to [31] and [1].

3.2   Galaxies as Weak Lenses

If the dark matter halos of individual galaxies are as massive as is suggested by dynamical [39] or hydrodynamical [8] arguments, they ought to act as extremely weak, yet detectable, gravitational lenses. That is, systematically throughout the universe, individual foreground galaxies should act as weak lenses for individual background galaxies. If one could detect this effect it would provide a probe of the gravitational potentials of the halos of the lens galaxies up to very large physical radii (on the order of $100h^{-1}$ kpc or so) where dynamical and hydrodynamical tracers of the potential are unlikely to be found. The advantage of using systematic weak lensing to probe dark matter halos is, therefore, that the method can be applied to all classes of galaxies. The disadvantage is that the weak lensing signal is so small that it cannot be detected convincingly for any one particular lens galaxy halo. That is, systematic galaxy–galaxy lensing only produces statistical answers.

Consider a large sample of galaxies which has been separated into a "foreground" and a "background" population. For foreground–background pairs which are relatively nearby to each other on the sky ($\theta \lesssim 100''$), the presence or absence of systematic lensing of the background population by the foreground population can be tested using the following simple statistic. For each background galaxy, the orientation ϕ of its equivalent image ellipse can be computed relative to the direction vector which connects its centroid to that of a foreground, and possible lens, galaxy (see Fig. 3.1). Using all pairs of candidate lenses and sources, then, the probability distribution of the orientation of a background galaxy with respect to the foreground galaxies, $P(\phi)$, can be computed. In the absence of systematic lensing of the background population by the foreground population, $P(\phi)$ will be consistent with a uniform distribution. However, if the background galaxies have been systematically distorted, there will be a very slight excess of pairs of galaxies in which the background galaxy is oriented tangentially, and a correspondingly slight deficit of pairs of galaxies in which the background galaxy is oriented radially. Specifically, if we introduce the two dimensional Newtonian potential Φ associated with a single lens galaxy at the location of a light ray from a distant source galaxy, we can introduce an optical potential,

$$\psi = \frac{2D_{LS}D_L}{c^2 D_S}\Phi \tag{3.1}$$

(where D_L, D_S, D_{LS} are the observer-lens, observer-source and lens-source angular diameter distances), and form the matrix

$$H_{ij} = \delta_{ij} - \frac{\partial^2 \psi}{\partial \theta_i \partial \theta_j}. \tag{3.2}$$

The optical shear associated with a ray congruence is defined by

$$\gamma = [\text{tr}(H)^2/4 - \det(H)]^{1/2} \tag{3.3}$$

which is clearly a property of the gravitational potential of the lens.

To first order, a circular isophote on a source galaxy lying behind a lens will be deformed into an ellipse with major and minor semi-axes, a, b. We can then define the image polarization p by

$$p = \frac{a^2 - b^2}{a^2 + b^2}. \tag{3.4}$$

Note that γ and $p(\boldsymbol{\theta}, z_s)$ can be made into complex scalar fields through multiplication by $\exp[2i\chi]$, where χ is the position angle on the sky. The polarization is, of course, not directly measured; however, a good statistical estimator of its value is furnished by averaging over a large number of source galaxies as we discuss below. Once an estimate of the polarization has been obtained, we can then extract the shear using the relation $\gamma = p/2$, which is valid in the linear regime. (The non-linear theory has been developed in several different ways, e.g., [31].)

In addition, we define a convergence,

$$\kappa = 1 - \text{tr}(H)/2 = \Sigma/\Sigma_{\text{crit}}, \tag{3.5}$$

where Σ is the surface density of the lens and

$$\Sigma_{\text{crit}} = \frac{D_S c^2}{4\pi G D_{LS} D_L}. \tag{3.6}$$

For a circularly symmetric mass distribution, $\gamma = \bar{\kappa} - \kappa$, where $\bar{\kappa}$ is the mean interior convergence. The associated image magnification is

$$\mu = \det[H^{-1}] = [(1 - \kappa)^2] - \gamma^2]^{-1} \tag{3.7}$$

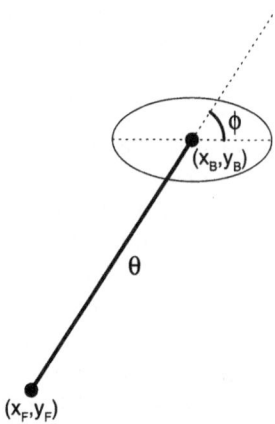

Fig. 3.1. Orientation, ϕ, of a background source galaxy relative to a foreground lens galaxy. Averaged over a large sample of galaxies in which galaxy–galaxy lensing has occurred, there will be a weak preference for background galaxies to be oriented tangentially (i.e., $\phi = 90°$)

When the distortion is small and we can use the linear approximation, the distribution of source galaxy position angles then becomes

$$P(\phi) = \frac{2}{\pi} \left[1 - \langle p \rangle \langle \epsilon^{-1} \rangle \cos 2\phi \right] \tag{3.8}$$

where $\langle p \rangle$ is the mean image polarization induced by gravitational lensing and $\langle \epsilon^{-1} \rangle$ is the harmonic mean of the (intrinsic) ellipticities of the weakly-lensed galaxies (see, e.g., [2]).

If the halos of the lens galaxies may be reasonably approximated as a singular isothermal spheres, the shear averaged within a circular annulus with inner radius θ_{in} and outer radius θ_{out} is given by

$$\langle \gamma \rangle = 4\pi \left(\frac{\sigma_v}{c} \right)^2 \left[\frac{D_{\text{LS}}}{D_{\text{S}}} \right] (\theta_{\text{in}} + \theta_{\text{out}})^{-1} \tag{3.9}$$

where σ_v is the velocity dispersion of the halo and c is the velocity of light. If the lens galaxy is a typical spiral at redshift $z_d \sim 0.5$ with $\sigma_v = 155$ km/s, then for source redshifts $z_s \sim 1$ we expect $\langle \gamma \rangle \simeq 0.007$ for $10'' \lesssim \theta \lesssim 30''$. This is not a level of shear which can be convincingly detected using existing data for any one particular lens galaxy, but in the limit of high–quality imaging data, an ensemble average over many thousands of candidate lens–source pairs should yield a statistically–significant detection of systematic galaxy–galaxy lensing.

3.3   Strategies for Detecting Galaxy–Galaxy Lensing

Detecting the systematic lensing of background galaxies by foreground galaxies is not trivial, owing to both the small size of the signal and the fact that no imaging system will ever be truly "perfect" (and, therefore, noise will be introduced into the images of the faint galaxies whose net distortion one is trying to determine). Wherever possible, of course, one tries to stack the deck in one's favor by using only the images of distant galaxies which are judged to have "reliable" shape information and by separating candidate lenses from candidate sources as best as possible. The latter requires a minimum of magnitude cuts on the data (i.e., on average, galaxies with "faint" apparent magnitudes will be located at greater distances from the observer than galaxies with "bright" apparent magnitudes); however, combining magnitude cuts with color information or photometric redshifts is certainly preferable whenever possible. In addition, a certain amount of care has to be taken to reject any candidate lens–source pairs in which the isophotes of the galaxies are close to overlapping since, in this case, galaxy detection algorithms could give spurious results for the shape of the fainter galaxy due to spillover light from the brighter galaxy.

There are two basic statistical approaches which investigators have taken in order to detect galaxy–galaxy lensing in their data sets: direct averaging of the signal and maximum likelihood. Both methods have their respective pros and cons, and below we briefly outline both methods.

3.3.1   Direct Averaging

This is an unsophisticated but very straightforward approach to the problem in which one simply computes the probability distribution of the observed position angles of the "background" galaxies (relative to the locations of "foreground" galaxies) and determines whether or not there is a preference for a tangential alignment of the background galaxies. If the background galaxies have been lensed systematically by the foreground galaxies, then as in (3.8) above the probability distribution of the position angles of the background galaxies, $P(\phi)$, should exhibit a $\cos 2\phi$ variation and the mean image polarization, $\langle p \rangle$, can be determined directly.

This technique is not particularly sophisticated from a statistical standpoint, but it is still a useful method because the significance of a detected signal does not depend upon the assumption of a prior (i.e., that lensing of the background galaxies by the foreground galaxies has, indeed, taken place). Rather, one simply tests the observed probability distribution against the null hypothesis (i.e., that the galaxies are not lensed and, therefore, are all randomly oriented), and the confidence level at which the null hypothesis is rejected is a measure of the significance of the detection.

Direct averaging also lends itself well to controlled null tests on the data by which systematic effects can be ruled out. In order to pass the null tests, $P(\phi)$ must be consistent with a random distribution. Null tests which have been incorporated in many of the recent galaxy–galaxy lensing investigations include:

- $P(\phi)$ for the orientation of the background galaxies, in which ϕ is computed relative to randomly-chosen locations on the sky. This tests for image shear induced by the telescope or the detector.
- $P(\phi)$ for the orientation of the background galaxies, in which ϕ is computed relative to bright foreground stars. This tests for effects due to diffraction spikes.
- $P(\phi)$ for the orientation of the foreground galaxies relative to the background galaxies (i.e., here ϕ is the orientation of the candidate lenses relative to the candidate sources). This checks the distance model.
- $P(\phi)$ for the orientation of the background galaxies after they have been rotated through 45°. This should be consistent with a uniform distribution and is a distinctive property of galaxy lensing.

Provided the data pass these null tests, one is can then be reasonably confident that any observed non–uniform distribution of $P(\phi)$ is, indeed, real and not an artifact of the data.

3.3.2   Maximum Likelihood

The direct averaging technique, while straightforward and better for studying the influence of systematic errors, is not the optimal statistical method and, in fact, it does not make use of all of the information which is present in the data since the observed shapes of the galaxies are not used when comparing the probability distribution of the image orientations to the null hypothesis. It is, therefore, not surprising that theoretical work has shown that a maximum likelihood approach to galaxy–galaxy lensing yields far better constraints on the characteristic physical parameters of the dark matter halos of the lens galaxies than does direct averaging [40].

This method takes into account the fact that weak gravitational lensing amounts to a change to an object's intrinsic ellipticity which is equal to the gravitational lensing shear itself. That is, if we define each galaxy's intrinsic shape to be

$$\epsilon^{(s)} = \frac{a^2 - b^2}{a^2 + b^2} e^{2i\phi} \simeq (1 - b/a)\, e^{2i\phi}, \tag{3.10}$$

where a and b are now the major and minor axes of the image and ϕ is its position angle, then to first order the relationship between the intrinsic ellipticity and ϵ, the observed shape after lensing, is given $\epsilon^{(s)} = \epsilon + \gamma$, where γ is the (complex) shear. The probability density for an observed galaxy shape is:

$$p_\epsilon(\epsilon|\gamma) = p^{(s)}\left(\epsilon + \gamma\right), \tag{3.11}$$

and an effective probability distribution can be obtained via many Monte Carlo simulations:

$$\langle p_i \rangle (\epsilon_i) = \frac{1}{N_{\mathrm{MC}}} \sum_{\nu=1}^{N_{\mathrm{MC}}} p^{(s)}(\epsilon_i + \gamma_{i,\nu}), \tag{3.12}$$

where N_{MC} is the number of Monte Carlo simulations. A likelihood function can be constructed from the probability densities:

$$\mathcal{L} = \prod_i \langle p_i \rangle (\epsilon_i) \tag{3.13}$$

and by varying the characteristic physical parameters of the lenses in the Monte Carlo simulations (e.g., the depths of potential wells and the characteristic radii), $\mathcal{L}$ can be maximized and the best-fit model parameters can be obtained (see, e.g., [40]).

In particular, Schneider & Rix [40] have demonstrated that if the redshifts of the candidate lens and source galaxies are known to within the accuracy which is typically obtained from photometric methods ($\Delta z \lesssim 0.1$), the characteristic extents of the dark matter halos of the lens galaxies should be constrained very well. As will be discussed further below, however, only one investigation to date [17] has actually been able to place direct constraints on the maximum extents of the dark matter halos of field galaxies with this method.

3.4   Detections of Galaxy–Galaxy Lensing

The first published attempt to detect systematic galaxy–galaxy lensing was that of Tyson et al. [44]. Despite a vast amount of data ($\sim 28,000$ foreground–background pairs), their result was consistent with a null detection on angular scales $\gtrsim 5''$. The first statistically–significant detection of galaxy–galaxy lensing to be published was the work done by Brainerd, Blandford & Smail [2], hereafter BBS, who used deep imaging data from a single CCD field (~ 72 sq. arcmin.) to investigate the orientation of 511 faint "background" galaxies relative to 439 brighter "foreground" galaxies. BBS claimed a formal 4-σ detection of galaxy–galaxy lensing on angular scales of $5'' \lesssim \theta \lesssim 35''$ and used their signal to place limits on the characteristic parameters of the dark matter halos of L^* field galaxies.

Since BBS, there have been 10 independent detections of galaxy–galaxy lensing by field galaxies ([7], [9], [12], [15], [17], [20], [21], [30], [43], [50]; see also Table 1). The data and its analysis vary considerably amongst these investigations, so it is difficult to compare all of the results directly. Not only does the imaging quality vary significantly, the categorization of the galaxies into "lenses" and "sources" is by no means consistent amongst the investigations. The latter is largely due to the fact that up until very recently, all of the galaxy–galaxy lensing detections were obtained with data which was not specifically acquired for the purposes of studying galaxy–galaxy lensing. Rather, the data are a broad heterogeneous mix of deep images which were oftentimes acquired for other purposes. Where possible, "lenses" were distinguished from "sources" on the basis of photometric redshifts, but most of the above investigations were limited to imaging in a single bandpass and, hence, only a very crude lens–source separation based upon apparent magnitude was performed (see Table 1). Nevertheless, the implications of these studies for the physical characteristics of the halos of field galaxies are all broadly consistent with one another.

To date, far and away the most statistically–significant detections of galaxy–galaxy lensing have come from analyses of the Sloan Digital Sky Survey (SDSS) commissioning data [12,30]. It is these results in particular which have made the study of galaxy–galaxy lensing much less controversial than it was a few years ago since they have demonstrated conclusively that even in the limit of somewhat poor imaging quality (including the presence of a mildly anisotropic point spread function), galaxy–galaxy lensing can be detected with very high significance in wide-field imaging surveys. In their most recent analysis of SDSS data, McKay et al. [30] have reported on observations of $\sim 3.6 \times 10^6$ source galaxies with $18 < r' < 22$ and $\sim 3.5 \times 10^4$ lens galaxies with $r' < 17.6$ for which spectroscopic redshifts were available. They obtain a measurement of the surface mass density contrast around lens galaxies which is well-fitted by a power law of the form $\Delta\Sigma_+ = \left(2.5^{+0.7}_{-0.6}\right) (R/1\mathrm{Mpc})^{-0.8\pm0.2} hM_\odot\mathrm{pc}^{-2}$ and investigate the dependence of galaxy–galaxy lensing on the luminosity and morphology of the lens galaxies, as well as the local environment. Although there are clear differences between the galaxy–galaxy lensing signals of spiral and elliptical lenses in this sample, a convenient statistic is the mass-to-light ratio within a fixed linear aperture of ~ 400 kpc. In the redder bandpasses, McKay et al. find M/L on this scale to be largely insensitive to the morphology of the lens galaxies, with a value of ~ 70 in solar units in both the i' and z' bands. A major concern about this measurement is, however, the degree to which it is influenced by intrinsic effects (see below).

3.4.1 Halo Model

Once galaxy–galaxy lensing has been detected at a convincing level, an obvious goal is to use the signal to constrain the characteristic parameters of the dark matter halos of the lens galaxies. To date, most of the investigations have followed the approach of BBS and have taken the halos of the galaxies to be modified singular isothermal spheres with a mass density which is given by

$$\rho(r) = \frac{\sigma_v^2 s^2}{2\pi G r^2 (r^2 + s^2)} \tag{3.14}$$

and for which the image polarization is

$$p(X) = \frac{4GMD_{\mathrm{L}}D_{\mathrm{LS}}}{s^2 D_{\mathrm{S}} c^2} \left[\frac{(2+X)(1+X^2)^{1/2} - (2+X^2)}{X^2(1+X^2)^{1/2}} \right], \tag{3.15}$$

where σ_v is the velocity dispersion of the halo, M is the total mass of the halo, $X \equiv R/s$, and R is the projected lens–source separation.

By assuming that a Tully–Fisher or Faber–Jackson type of relation will hold for the galaxies, and also that the total mass–to–light ratio is constant independent of luminosity, it is then possible to relate the velocity dispersion which is associated with a given halo (σ_v), the luminosity of the galaxy which resides within the halo (L), and the characteristic outer radius of the halo (s), to these same parameters for an L^* galaxy via simple scaling relations:

$$\frac{\sigma_v}{\sigma_v^*} = \left(\frac{L}{L^*}\right)^{1/4} \qquad \frac{s}{s^*} = \left(\frac{L}{L^*}\right)^{1/2} \qquad (3.16)$$

(see, e.g., [2]). Monte Carlo simulations of galaxy–galaxy lensing can then be used to determine the values of σ_v^* and s^* which best reproduce the observed signal in the data using either a χ^2 or maximum likelihood approach.

Table 1 summarizes the current published (or soon to be published) detections of galaxy–galaxy lensing by field galaxies, along with the characteristic physical parameters which have been inferred for the halos of L^* galaxies. Perhaps surprisingly, in spite of the substantial variety in both the data and the analysis techniques, there is remarkable agreement amongst these studies. All obtain "reasonable" velocity dispersions for the halos of L^* galaxies (in the range of 130 km/s to 190 km/s) and all but one have been unable to place a direct constraint on the *maximum* radial extent of the halos of L^* field galaxies, s^*. The exception to this is Hoekstra's result [17] which gives a 1-σ upper bound of order $600h^{-1}$ kpc and a 2-σ upper bound of order $1h^{-1}$ Mpc for the radius of the halos of L^* field galaxies.

The primary difficulty with using galaxy–galaxy lensing to constrain the characteristic outer scale radii of the halos is simply due to the fact that the signal is relatively insensitive to this parameter (unlike the velocity dispersion, to which it is quite sensitive). Nevertheless, it is expected that considerably larger data sets with imaging quality similar to [17] should allow this situation to improve.

3.5   Applications of Galaxy–Galaxy Lensing

We have not yet learned much that is both new and secure from the recent detections of galaxy–galaxy lensing. That is, all of the constraints on the nature of dark matter halos which have been obtained from galaxy–galaxy lensing agree well with previous constraints which were obtained by more conventional methods. These studies have, however, demonstrated quite convincingly that galaxy–galaxy lensing is a viable technique by which the dark matter distribution on the scales of individual galaxies (i.e., proper distances of order a few 100 kpc) may be investigated.

Despite the fact that it is challenging to detect, galaxy–galaxy lensing holds the promise of becoming a technique by which fundamental questions about the history of galaxy formation can be addressed directly. It is expected that, at least in principle, high signal–to–noise observations of galaxy–galaxy lensing could provide some unique constraints on galaxy formation which are unlikely, if not impossible, to obtain with other, more traditional techniques. It should, however, be noted that because galaxy–galaxy lensing occurs in the weak lensing regime and it is unaffected by the presence or absence of reasonably–sized core

Table 1: Detections of Galaxy–Galaxy Lensing by Field Galaxies

	Data	σ_v^*	radius?
Brainerd, Blandford & Smail (1996)	72 sq. arcmin. $20 \leq r_{\mathrm{bright}} \leq 23$ (439 obj) $23 < r_{\mathrm{faint}} \leq 24$ (511 obj.) 3202 faint-bright pairs	155 ± 56 km/s	$s_* \gtrsim 100\ h^{-1}$ kpc
Griffiths et al. (1996)	HST Medium Deep Survey + $15 < I_{\mathrm{bright}} < 22$ (1600 obj.) $22 < I_{\mathrm{faint}} < 26$ (14,000 obj.)	220 km/s (E) 155 km/s (S)	$r_t \simeq 100 r_{hl}$
Dell'Antonio & Tyson (1996)	HDF North $22 < I < 25$ + colour cut 110 lenses, 645 BLUE sources	185^{+30}_{-35} km/s	$r_{\mathrm{out}} \geq 15 h^{-1}$ kpc
Hudson et al. (1998)	HDF North photo-z's, $0 \leq z \leq 0.85$ 208 lenses, 697 sources	148 ± 28 km/s	no strong constraint
Ebbels (1998)	HST Medium Deep Survey + $18 < I_{\mathrm{bright}} < 22.5$ $22.5 < I_{\mathrm{faint}} < 25.5$ 22,000 objects	128^{+25}_{-34} km/s	$r_t^* > 120$ kpc
Fischer et al. (2000)	SDSS commissioning data 225 sq. degrees $16 \leq r'_{\mathrm{bright}} \leq 18$ (28,000 obj.) $18 \leq r'_{\mathrm{faint}} \leq 22$ (150,000 obj.)	145 – 195 km/s	$s_* > 275 h^{-1}$ kpc
Hoekstra (2000)	CNOC-2 fields $17.5 < R_{\mathrm{bright}} < 23$ $22 < R_{\mathrm{faint}} < 26$	133^{+14}_{-15} km/s	$s_* = 260^{+124}_{-73} h^{-1}$ kpc
Jaunsen (2000)	CFRS 14-hr field $UBVR_CI_C$, photo-z	141^{36}_{53} km/s	no constraint
McKay et al. (2001)	SDSS (imaging + spectra) $r'_{\mathrm{lens}} < 17.6 + z_{\mathrm{spec}}$ (3.4×10^4 obj.) $18 < r' < 22$ (3.6×10^6 obj.)	100 – 130 km/s	$s > 230 h^{-1}$ kpc
Smith et al. (2001)	LCRS lenses $R_{\mathrm{L}} < 18 + z_{\mathrm{spec}}$ (790 obj.) $20 < R_{\mathrm{S}} < 23$ (450,000 obj.)	116^{+14}_{-14} km/s	no constraint
Wilson et al. (2001)	photo-z's, EARLY type lenses $0.1 \leq z_{\mathrm{lens}} \leq 0.9$ (15,000 obj.) $I_{\mathrm{source}} > 25$ (148,000 obj.)	168^{+19}_{-21} km/s	no constraint

radii within the lens galaxies, galaxy–galaxy lensing will not help to resolve the current debate over the cuspiness of galaxy halos as predicted by the Cold Dark Matter scenario (e.g., [36,32]).

Primary issues for which galaxy–galaxy lensing should provide particularly useful statistical constraints are:

- the typical physical parameters which are associated with the dark matter halos of galaxies, including any systematic deviations of the halos from pure spherical symmetry (i.e., flattened halos)
- the degree to which the dark matter halos of galaxies are truncated during the infall of galaxies into cluster environments
- the morphological dependence of the halo potential (i.e., early– versus late–type galaxies)
- the evolution of the total mass–to–light ratio of galaxies, both in the field and in clusters
- the scaling of total galaxy mass with luminosity (i.e., $M \propto L^\alpha$), including any strong evolution of the Tully–Fisher and Faber–Jackson relations with redshift
- the "bias" of light versus mass in the universe via the galaxy–mass correlation function
- the shape of the redshift distribution of distant, faint galaxies whose redshifts fall between $z \sim 1$ and $z \sim 3$

Both observational and theoretical investigations into the use of galaxy–galaxy lensing to address all of the above issues have begun, although at present the constraints which have been obtained are not especially strong. The preliminary results are, however, sufficiently interesting to justify significantly more work in the future and below we summarize a few of the current investigations.

3.5.1   Flattened Galaxy Halos

Although it is true that the simple, singular, isothermal sphere can reproduce the flatness of the rotation curves of the disks of spiral galaxies, there are both observational and theoretical arguments in favor of halos which are flattened, rather than spherical. The observational evidence is somewhat scarce, owing to the fact that there are relatively few galaxies for which the shape of the halo potential can be probed directly via traditional methods. Nevertheless, the evidence for flattened halos is quite diverse and includes such observations as the dynamics of polar ring galaxies, the geometry of X-ray isophotes, the flaring of HI gas in spirals, the evolution of gaseous warps, and the kinematics of Population II stars. In particular, studies of disk systems which probe distances of order 15 kpc from the galactic planes suggest that the ratio of shortest to longest principle axes of the halos is $c/a = 0.5 \pm 0.2$ (see, e.g., [38] and references therein). Additionally, some instances of strong lensing by individual galaxies suggest that the halos of the lenses are not spherical. For example, provided the disk mass is small compared to the mass of the halo, then the halo of the spiral galaxy which lenses the quasar B1600+434 may have a value of c/a that is as low as 0.53, [28]. In addition, high-resolution simulations of dissipationless cold dark matter models consistently result in triaxial (not spherical) galaxy halos with a median ellipticity of order 0.3 (see, e.g., [8] and [47]). Therefore, from a theoretical standpoint we would expect that the halos will be flattened in projection.

Galaxy–galaxy lensing has the potential to provide constraints on the mean flattening of the dark matter halos of field galaxies and recently two groups have investigated galaxy–galaxy lensing by flattened halos in order to assess the amount of data which would be required in order to detect the effects of flattened halos on the lensing signal [3,35]. For simplicity, both groups modeled the dark matter halos as infinite singular isothermal ellipsoids with identical ellipticities (i.e., a range of halo ellipticities was not included) and it was assumed that each distant source galaxy would only be lensed by a single foreground galaxy (i.e., multiple deflections were not taken into consideration). In addition, it was assumed that the position angle of the major axis of the *mass distribution* of the lens galaxies will be fairly well–aligned with the (unlensed) major axis of the *light distribution*. Provided a galaxy is in a state of dynamical equilibrium (i.e., it has not undergone a recent collision or merger) the latter is, of course, a reasonable expectation.

Gravitational lensing by an elliptical halo gives rise to a shear pattern which is anisotropic about the lens center such that at a given angular distance, θ, from the lens center (and at fixed source redshift, z_s), the magnitude of the shear is greatest for sources located nearest to the major axis of the lens and least for sources located nearest to the minor axis of the lens. Hence, in a given radial annulus which is centered on the lens, the mean shear experienced by sources whose azimuthal coordinate, φ, places them within $\pm N°$ of the major axis of the lens will be greater than that for sources whose azimuthal coordinate, φ, places them within $\pm N°$ of the minor axis. As a shorthand notation, we will refer to the magnitude of the mean shear experienced by sources whose azimuthal coordinates place them within $\pm N°$ of the minor axis of the lens as $\langle \gamma^- \rangle$. Similarly, we will refer to the magnitude of the mean shear experienced by sources whose azimuthal coordinates place them within $\pm N°$ of the minor axis of the lens as $\langle \gamma^+ \rangle$ (see, e.g., Fig. 3.2).

To estimate roughly the size of a survey which would be needed to detect anisotropic galaxy–galaxy lensing, let us begin by considering a completely isolated, singular isothermal ellipsoid lens for which the convergence is:

$$\kappa(r, \varphi) = \frac{\sqrt{f}}{2r\Delta(\varphi)}, \tag{3.17}$$

where $\Delta(\varphi) = \sqrt{\cos^2\varphi + f^2 \sin^2\varphi}$, $f = b/a$ is the axis ratio of the mass distribution $(0 < f \leq 1)$, and r is a radius vector projected on the sky (see, e.g. [26]). The components of the complex shear, $\gamma = \gamma_1 + i\gamma_2$, are given by:

$$\gamma_1 = -\kappa \cos 2\varphi \tag{3.18}$$

$$\gamma_2 = -\kappa \sin 2\varphi \tag{3.19}$$

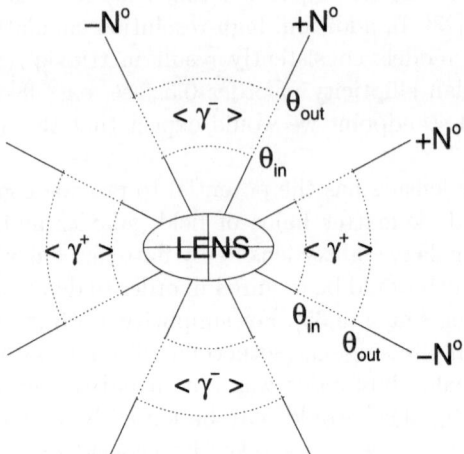

Fig. 3.2. Illustration of the anisotropic shear field about a lens with an ellipsoidal mass distribution. At a given radial distance from the lens center, θ, (and at fixed source redshift z_s), sources which are located closest to the major axes of the lens will experience greater shear than those which are located closest to the minor axes. Figure courtesy of C. O. Wright.

and the modulus of the shear is $\gamma = \sqrt{\gamma_1^2 + \gamma_2^2} = \kappa$, in direct analogy with the singular isothermal sphere lens.

If we compute $\langle \gamma^+ \rangle$ and $\langle \gamma^- \rangle$ for all sources whose azimuthal coordinates, φ, are within $\pm N°$ of the symmetry axes of such a lens, it can be shown that at any angular radius, θ, the following relationship will hold:

$$\frac{\langle \gamma^- \rangle}{\langle \gamma^+ \rangle} = \frac{\left\langle \frac{1}{\Delta(\varphi)}^- \right\rangle}{\left\langle \frac{1}{\Delta(\varphi)}^+ \right\rangle}. \tag{3.20}$$

The ratio $\langle \gamma^- \rangle / \langle \gamma^+ \rangle$ is shown in the left panel of Fig. 3.3 for lens mass ellipticities in the range 0 to 0.7 (i.e., values of f in the range 1 to 0.3) and for $N = 10°, 20°, 30°, 40°, 45°$. For "reasonable" halo mass ellipticities (i.e., $\epsilon \sim 0.3$) the shear ratio is considerably smaller than a factor of 2 but, nevertheless, even such a relatively small anisotropy is potentially measurable with an appropriate data set.

If we next define an anisotropy parameter to be

$$\mathcal{A} = 1 - \left[\langle \gamma^- \rangle / \langle \gamma^+ \rangle \right], \tag{3.21}$$

it is then straightforward to show that in order to obtain an $M\sigma$ detection of $\mathcal{A}$, the signal–to–noise in the measurements of $\langle \gamma^+ \rangle$ and $\langle \gamma^- \rangle$ would each need to be of order

$$\frac{\sqrt{2}M \langle \gamma^- \rangle}{\langle \gamma^+ \rangle - \langle \gamma^- \rangle}. \tag{3.22}$$

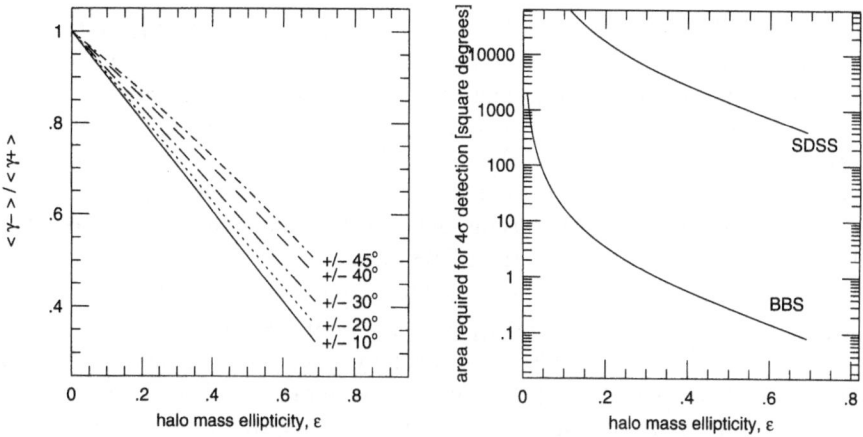

Fig. 3.3. Left panel: Anisotropy in galaxy-galaxy lensing due to flattened halos, expressed as a ratio of the shear experienced by sources located closest to the minor axis of the lens to that experienced by sources located closest to the major axis of the lens. The lenses have been modeled as untruncated singular isothermal ellipsoids. Results are shown for sources which have polar angles (relative to the symmetry axes of the lens) of $\pm 45°$, $\pm 40°$, $\pm 30°$, $\pm 30°$, and $\pm 10°$. Right panel: Survey area expected for a 4-σ detection of anisotropic galaxy–galaxy lensing as a function of the ellipticity of the halos. Lens–source separations of $\lesssim 40''$ have been adopted and the S/N obtained in the original BBS and SDSS data sets have been used in the estimate. Note that this is somewhat a pessimistic prediction which assumes that the S/N in the final SDSS data will not be significantly improved compared to the commissioning data. For this panel we have computed the anisotropy parameter, $\mathcal{A}$, using an unweighted mean and all sources within $\pm 45°$ of the symmetry axes of the lenses.

(see, e.g., [3]). Additionally, if all else is equal (i.e., the depth, seeing, and noise properties of the imaging data), the signal–to–noise in a detection of galaxy–galaxy lensing scales as the square root of the area of the data set (see, e.g., [2], [35]). That is, if two data sets have identical imaging characteristics but differ by a factor of 4 in the area of sky that is covered, the larger survey will yield a signal–to–noise in a detection of galaxy–galaxy lensing that is a factor of 2 larger than would be yielded by the smaller survey.

Knowing the above, one can then ask the question: How large an area would previous investigations have had to have covered in order to detect anisotropic galaxy–galaxy lensing in their data sets ? As an illustrative example, the right panel of Fig. 3.3 shows the area of sky which would have been needed by the relatively deep imaging of BBS ($r \leq 24$) and also by the relatively shallow imaging of SDSS ($r \leq 22$) in order to obtain a 4-σ detection of anisotropic galaxy–galaxy lensing in their data sets, provided all pairs of lenses and sources which were separated by $\theta \lesssim 40''$ were used in the analysis. Again, note that here it has been assumed that all halos have identical ellipticities in projection on the sky and that multiple deflections have only a negligible effect on the final image shapes of the sources. Note also that the signal–to–noise in the SDSS data was taken to be identical to the rather poor signal–to–noise in the commissioning data used by Fischer et al. [12]. This should, of course, change considerably in the final SDSS data set since it certainly hoped that the imaging quality will be greatly improved and, of course, the distinction between lenses and sources will be based upon spectroscopic redshifts (rather than apparent magnitudes as in the original analysis, [12]). The right panel of Fig. 3.3, therefore, represents a rather pessimistic prediction for the SDSS data.

Given that we expect the mean halo ellipticity to be of order 0.3 in projection (based upon theoretical arguments), Fig. 3.3 would certainly seem to suggest that the effects of flattened halos on the galaxy–galaxy lensing signal should be detectable with a modestly deep, relatively wide data set with good imaging (see also [3]). In addition to the estimates shown in Fig. 3.3, the analytical calculation done by Natarajan & Refregier [35] for singular isothermal ellipsoid halos yields an estimate for the signal–to–noise in a measurement of the ellipticity of the halo mass, $\epsilon_\kappa \equiv (a^2 - b^2)/(a^2 + b^2)$, which is given by

$$\left(\frac{S}{N}\right)_{\epsilon_\kappa} \simeq 4.6 \left(\frac{\epsilon_\kappa}{0.3}\right) \left(\frac{\alpha}{0.5''}\right) \left(\frac{n_b}{1.5}\right)^{1/2} \left(\frac{n_f}{0.035}\right)^{1/2} \left(\frac{0.3}{\sigma_\epsilon}\right) \left(\frac{A}{1000}\right)^{1/2}. \quad (3.23)$$

Here α is the Einstein radius, n_b is the number of background galaxies per square arcminute, n_f is the number of foreground galaxies per square arcminute, σ_ϵ^2 is the variance in the intrinsic ellipticity distribution of galaxies ($\sim 0.3^2$), and A is the area of the survey in square degrees. Similarly to the right panel of Fig. 3.3, (3.23) explicitly assumes that all halos have identical ellipticities in projection and that multiple deflections do not have a significant effect on the weak lensing signal. For halos with ellipticities of order 0.3 (i.e., axis ratios, f, of 0.7), (3.23) predicts that a data set similar to the BBS data set would require of order 6 square degrees to detect the effects of flattened halos on the galaxy–galaxy lensing signal at a 4-σ level. In the case of a data set similar to the SDSS data set, it predicts that of order 600 square degrees would be needed to achieve a 4-σ detection. It should be noted, however, that (3.23) accounts for only one source of noise in the data, the intrinsic ellipticity distribution of the source galaxies, and is explicitly independent of the lens–source separation on the sky. Therefore, it is unsurprising that, although they are based upon identical halo mass models, the predictions for the sizes of surveys which would be needed in order to detect anisotropic galaxy–galaxy lensing using (3.23) and the right panel of Fig. 3.3 agree only to within an order of magnitude (see, e.g., [3]).

It can certainly be argued that, at least in the case of late–type lens galaxies, the effect of multiple deflections (i.e., source galaxies being weakly lensed at comparable levels by two or more foreground galaxies) should be minimal in a shallow data set such as the SDSS. However, for deeper data sets, such at that used by BBS, the effect could be quite substantial and, in fact, the Monte Carlo simulations performed by BBS indicated that of order one third of the galaxies with magnitudes in the range $23 \leq r \leq 24$ would have been lensed at a comparable level by two foreground galaxies, and another third of the galaxies would have been lensed at a comparable level by three or four foreground galaxies.

In order to improve estimates of the size of a deep imaging data set which would be required to detect the effects of halo flattening on the galaxy–galaxy lensing signal, Wright [52] has been performing detailed Monte Carlo simulations of galaxy–galaxy lensing by flattened halos. Her simulations attempt to reproduce a number of observational constraints on the faint galaxy population, such as the number counts of faint galaxies, $\frac{d \log N}{dm}$, to a limiting magnitude of $I_{\text{lim}} = 25$, the shape of the redshift distribution of faint galaxies, $N(z)$, (extrapolated to $I_{\text{lim}} = 25$), and the distribution of intrinsic image shapes (i.e., the unlensed equivalent ellipses of the light distribution of the sources), as obtained from deep imaging with HST. The halos are modeled as truncated singular isothermal ellipsoids with surface mass densities given by:

$$\Sigma(\rho) = \frac{\sigma_v^2 \sqrt{f}}{2G} \left(\frac{1}{\rho} - \frac{1}{\sqrt{\rho^2 + x_t^2}} \right), \tag{3.24}$$

where σ_v is the velocity dispersion, f is the axis ratio of the mass distribution as projected on the sky ($0 < f \leq 1$), x_t is a truncation radius, G is Newton's constant, and ρ is a generalized elliptical radius defined such that $\rho^2 = x_1^2 + f^2 x_2^2$. Here x_1 and x_2 are Cartesian coordinates measured, respectively, along the minor and major axes of the projected mass distribution of the halo (see, e.g., [26]). The ellipticities of the halos are drawn from a probability distribution which is based upon current observational constraints (i.e., from principle moment analyses in which, by definition, $a > b > c$, the distribution of halo shapes seems to favor $c/a = 0.5 \pm 0.2$ and $b/a \gtrsim 0.8$; see [38]). The distribution has both a mean and a median halo ellipticity which are of order 0.3. Similarly to BBS, it is also assumed that a Tully–Fisher or Faber–Jackson type law will hold for the lenses and that the mass–to–light ratio of the galaxies is constant independent of its luminosity, in which case the velocity dispersions and truncation radii of the halos can be scaled in terms of a fiducial velocity dispersion and truncation radius of an L^* galaxy.

Wright's Monte Carlo simulations of galaxy–galaxy lensing by flattened halos take full account of multiple deflections for all galaxies and have been used to compute the area of a survey which would be needed to detect the anisotropy parameter, $\mathcal{A}$, at a significant level. Shown in the left panel of Fig. 3.4 is the signal–to–noise in a measurement of $\mathcal{A}$ as a function of the survey area for lenses and sources which are separated by $5'' \leq \theta \leq 35''$ on the sky. Only galaxies with apparent magnitudes in the range $19 \leq I \leq 23$ (i.e., very similar to the original BBS data set) have been used in the calculation. In the left panel no attempt has been made to include a reasonable level of observational noise in the calculation; instead, the actual values of the redshift, position angle of the halo, and position angle of the image (after lensing) have been used in the calculation. An unweighted average is used to compute $\langle \gamma^- \rangle$ and $\langle \gamma^+ \rangle$ for sources within $\pm 20°$ and $\pm 45°$ of the symmetry axes of the halos of the lens galaxies (see Fig. 3.2). Note that, because the anisotropy increases with decreasing N (i.e., the left

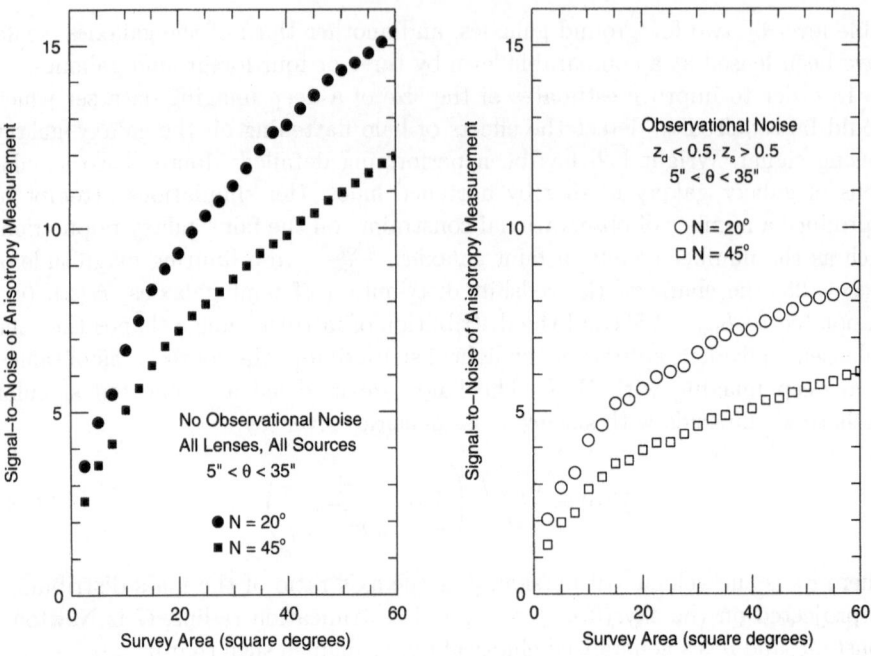

Fig. 3.4. Left panel: Signal-to-noise in a measurement of $\mathcal{A}$ using all galaxies with $19 \leq I \leq 23$ in the Monte Carlo simulations which are separated by $5'' \leq \theta \leq 35''$ on the sky. Complete information (exact values of redshift, position angle of the image, and position angle of the halo) is used. Unweighted averages are used to compute the mean shear for sources with azimuthal coordinates which place them within $\pm 20°$ (circles) and $\pm 45°$ (squares) of the symmetry axes of the lenses. Right panel: Same as left panel, but here observational noise has been added to the data. The analysis is restricted to lenses with redshifts $z_d < 0.5$ and sources with $z_s > 1.0$ (see text). Figure courtesy of C. O. Wright.

panel of Fig. 3.3), a somewhat greater value of the signal–to–noise is achieved for $N = 20°$ than for $N = 45°$ for a survey of a given area.

The left panel of Fig. 3.4 yields a prediction for the size of a BBS–like data set which would be needed to detect the effects of flattened halos. It differs from the right panel of Fig. 3.3 in that: (1) a broad distribution of halo ellipticities, redshifts, and velocity dispersions has been used, (2) multiple deflections have been accounted for, and (3) no observational noise has been included. The left panel of Fig. 3.4 suggests that only a relatively modest amount of data (about 3.5 square degrees for $N = 20°$ and about 6.25 square degrees for $N = 45°$) would be needed to detect anisotropic galaxy–galaxy lensing at a 4-σ level. In the limit of realistic data, however, this turns out to be an overly–optimistic estimate of the requisite size of the data set.

One of the fundamental assumptions which went into the first estimates of the detectability of anisotropic galaxy–galaxy lensing [3,35] was that the orientation of the image of a lens galaxy would be completely uncorrelated with the orientation of the image of a source galaxy. However, as is well–known from studies of lensing by large–scale structure [45], the orientations of galaxies which are nearby to each other on sky will be weakly–correlated. In her simulations, Wright [52] has found that the correlation is sufficiently strong that, unless she divides her sample into foreground galaxies with $z_d < 0.5$ and background galaxies with $z_s > 0.5$, the anisotropy in the galaxy–galaxy lensing signal due to flattened halos is, for all intents and purposes, unobservable. This is due to the fact that a single lens at redshift z_1 will induce correlated ellipticities are images of a lens–source pair whose redshifts are z_2 and z_3 (here $z_1 < z_2 < z_3$). Hence, the orientation of the major axis of the *image* of a foreground galaxy will have been systematically misaligned away from the orientation of the *halo*, as projected on the sky.

When Wright [52] accounts for this effect and, additionally, includes a realistic level of observational noise in her simulations, she finds survey areas which are substantially larger than the early estimates are needed to detect the effects of anisotropic galaxy–galaxy lensing (right panel of Fig. 3.4). Noise which has been added includes errors in the redshift which are typical of photometric estimates (i.e., $\Delta z_{\mathrm{phot}} \lesssim 0.1$) and errors in the image shapes caused by seeing, pixellation, sky noise, etc. It would appear from this work that the signal will be detectable in the long run, but that it will require a substantial amount of data (of order 20 to 40 square degrees) for a significant detection.

3.5.2   Galaxy–Galaxy Lensing Through Clusters

The systematic weak lensing of background galaxies by the individual galaxies within lensing clusters has been studied in two published manuscripts to date [13,33]. That is, these investigations have searched for instances of galaxy–galaxy lensing in which the lens galaxies are embedded within a larger cluster potential. Although at the outset it might seem impossible to detect the galaxy–galaxy lensing signal due to the much larger cluster potential, it is actually not much more difficult than detecting galaxy–galaxy lensing in the field. This is, in part, due to the fact that on the angular scales over which galaxy–galaxy lensing is detected (a few arcseconds or so), the cluster potential changes very slowly and, so, weak lensing by the cluster does not swamp the galaxy–galaxy lensing signal. In fact, conclusions about the physical properties of the individual halos of the cluster galaxies are only weakly–dependent upon the removal of the larger cluster potential.

In addition, since lensing conserves surface brightness and since cluster lenses magnify the images of distant galaxies somewhat, it is possible to use sources which are more distant than one would otherwise be able to use in the statistical analysis. The primary motivation for such work is, of course, to determine whether or not the dark matter halos of cluster galaxies have been truncated significantly compared to those of field galaxies, and the preliminary results from some of the current investigations seem quite promising.

Both [13,33] used HST WFPC2 data for their investigations and both successfully detected galaxy–galaxy lensing within the cluster fields (Cl0939+4713 in the case of [13] and AC114 in the case of [33]). Despite a detection of galaxy–galaxy lensing in Cl0939+4713, however, the field appeared to be too small to allow strong conclusions to be drawn about the mass distributions of the cluster galaxies. In the case of AC114, on the other hand, four WFPC2 pointings were used to construct a wide–field mosaic and the subsequent detection of galaxy–galaxy lensing led to the conclusion that the halos of the cluster galaxies are substantially smaller in radial extent and, correspondingly, are less massive than the halos of field galaxies.

Recently, Natarajan et al. have extended their initial work to include a set of five galaxy clusters which span a wide range of redshifts ($0.18 \leq z \leq 0.58$) and they find that not only are the dark matter halos of the cluster galaxies truncated compared to those of field galaxies, the proper length of the truncation radius increases with the redshift of the cluster and the total mass–to–light ratio of the cluster galaxies also increases with the redshift of the cluster [34]. This is, of course, precisely the behavior which one would expect based upon theoretical grounds (e.g., [14]) and, therefore, this is potentially a landmark result. However, there are a few nagging concerns which should be kept in mind:

- The clusters not only span a wide range of redshift, they also span a wide range in richness, mass, and X–ray luminosity (i.e., they are a very heterogeneous sample).
- It is expected that the higher the redshift of the cluster, the greater is the likelihood of the cluster containing an increased level of unresolved substructure, which could be attributed erroneously to the cluster galaxies. (This is in part due to the fact that the higher redshift clusters contain fewer strongly–lensed sources, which are an important element in constraining the overall cluster mass model.)
- The depth of the line of sight to the more distant clusters makes it increasingly likely that there will be an increased proportion of contaminating field galaxies.
- Differential luminosity evolution could bias the selection towards intrinsically less massive galaxies at low redshift.

Nevertheless, if the systematic errors can be controlled at a sufficient level, galaxy–galaxy lensing through clusters should, in the long run, be able to provide a unique insight into the assembly history of clusters and perhaps even yield observational constraints with which theory can be tested.

3.5.3   Morphological Dependence of the Halo Potential

Both dynamical studies and the fact that early–type galaxies are more commonly observed to be strong lenses than are late–type galaxies (e.g., [25]) suggest that the depth of the potential wells of early–type L^* galaxies should be significantly deeper than those of late–type L^* galaxies. If this is the case, it should be detectable in studies of galaxy–galaxy lensing for which morphological information is available and, hence, one should be able to use galaxy–galaxy lensing to constrain the mean halo velocity dispersion as a function of morphology.

The results to date have been somewhat mixed, but are generally encouraging. In their original study of galaxy–galaxy lensing in the Medium Deep Survey fields, [15] split their sample into early–types and late–types based upon the visual morphology. They clearly found that the galaxy–galaxy lensing signal was stronger for their early–type galaxies than for their late–type galaxies, and they inferred correspondingly larger velocity dispersions for the halos of the early–type galaxies (see Table 1). Wilson et al. [50] restricted their galaxy–galaxy lensing analysis to only those galaxies whose $(V - I)$ colors were consistent with an early–type population, in part because they found that if they did not make this restriction they were unable to detect the lensing signal at all [51]. However, Wilson et al. inferred a velocity dispersion for their "early–type" L^* galaxies which is only marginally larger than the velocity dispersion inferred by others who did not attempt to split the data on lens morphology.

More recently, the SDSS group has been investigating the morphological dependence of galaxy–galaxy lensing, and have the great luxury of being able to make the morphological discrimination based upon the spectra of the lenses. As was the case with the original MDS result, the SDSS group finds a substantially stronger galaxy–galaxy lensing signal is exhibited by the early–type galaxies than the late–type galaxies (see, e.g., Fig. 16 in [30]). Their data now cover at least four times the area used in their original analysis, and their preliminary work with this larger data set suggests that within a proper radius of 250 kpc, the halos of early–type L_* galaxies are 2.5 times more massive than the halos of late–type L_* galaxies [29].

Although in some sense heartening, these preliminary results should, of course, be assessed with some degree of caution because of the strong dependence of galaxy morphology on local density (i.e., morphology segregation; [27]). The recent analysis of the SDSS data by McKay et al. [30] does, indeed, indicate a dependence of the galaxy–galaxy lensing signal on local environment (i.e., lenses in "low" versus "high" density regions as determined from a Voroni tesselation). The effect of the local environment in their data appears to be relatively small on scales $\lesssim 200$ kpc (see, e.g., Fig. 17 in [30]) but, nevertheless, until such time as any contribution to the inferred value of σ_v^* for ellipticals which might be caused by the presence of local clusters is eliminated, the true sizes of the error bars on σ_v^* will remain somewhat in doubt.

3.5.4 Bias Factor

It has long been suspected that the clustering of galaxies (i.e., the light in the universe) is biased relative to that of the underlying mass distribution such that the autocorrelation functions of the galaxies and the mass are related through:

$$\xi_{gg}(r) = b^2(r)\xi_{mm}(r) \tag{3.25}$$

where ξ_{gg} is the galaxy autocorrelation function, ξ_{mm} is the autocorrelation function of the mass, and b is a "bias factor" which, in general, may be scale-dependent [22]. Weak lensing of the galaxy population provides a direct measurement of this bias since the variation of the tangential shear with angular separation is, in effect, a cross-correlation between the galaxies and the mass distribution of the universe [23]. In particular, since the galaxy-galaxy correlation function is well-fit by a power law [27], and the variation of the tangential shear from galaxy-galaxy lensing is also well-fit by a power law, $\gamma(\theta) \propto \theta^{-\eta}$, [12], the galaxy-mass correlation function will also be a power law:

$$\xi_{gm}(r) = \left(\frac{r_{gm}}{r}\right)^{\gamma} \tag{3.26}$$

where $\gamma = \eta + 1$. Combining the Limber's equation for $\gamma(\theta)$ and the observed power-law dependence of the galaxy autocorrelation function yields a measurement of r_{gm} [23]. Arriving at a specific value of r_{gm} is, of course, somewhat complicated by the fact that the correlation length, r_0, of the local galaxy autocorrelation function is strongly affected by both the morphology of the galaxies in the sample (i.e., the morphology–density relation) and by the intrinsic luminosities of the galaxies as well (i.e., luminosity segregation; see e.g., [27]).

A preliminary estimate of $r_{gm} \sim 3h^{-1}\Omega_{m0}^{-0.57}$ Mpc was obtained by Fischer et al. [12] from the analysis of galaxy–galaxy lensing in the SDSS commissioning data. More recently, Hoekstra et al. [18] used data from the Red-Sequence Cluster Survey to obtain a measurement of b/r, where r is the galaxy–mass correlation coefficient. The value of b/r is dependent upon the cosmology, and for the currently popular flat, lambda-dominated model, Hoekstra et al. find $b/r = 1.05^{+0.12}_{-0.10}$. Although neither of these measurements is definitive, they certainly represent encouraging progress.

3.5.5   Lensing of Halos

Much of the motivation for pursuing investigations of galaxy–galaxy lensing has been to learn about the lenses, but it is our opinion that it is equally interesting to use galaxy–galaxy lensing to learn about the *sources*. Because of this, we now turn to a rather different example of weak lensing which can tell us about dark halos through their role as sources. When one counts sources on the deepest HST fields that have been observed, it is found that there are roughly 100 billion faint sources on the sky with magnitudes $I \sim 30$ (see, e.g., [49]). It is known that these sources. are blue and extremely compact (e.g., [41]), or at least centrally concentrated; however, their location, properties, provenance, and fate are a mystery. Presuming, on the basis of their color, that they mostly have redshifts less than $z \sim 3$ their comoving density is roughly an order of magnitude greater than that of local bright galaxies. Their most natural interpretation is that they are protogalaxies that subsequently combine to produce the galaxies that we see around us today. However, other hypotheses have been entertained. They could be a population of extremely low luminosity galaxies with $z << 1$; they could be a population of dwarf galaxies with $z \sim 1$ from which the interstellar medium has been expelled by the first supernovae so that they quickly fade from view; they could be satellites of large, nearby galaxies. Undoubtedly, these explanations are all true at some level. However, we shall assume that the majority of these faint sources are the building blocks of contemporary galaxies and we are consequently very interested to learn their redshift distribution. We could then compute their luminosities and the physical scale of their clustering. We could also determine their rest-frame colors and then estimate their star formation rates. In particular it is of interest to ask if they are better characterized as discrete star-forming regions in relaxed, galaxy-size halos or discrete protogalaxies, each carrying small halos that have yet to merge. Answering these questions would tell us much of empirical value about galaxy formation.

These sources are far too faint, by 3-5 magnitudes, for direct spectroscopy. They are also too small for a conventional, weak lensing determination of their redshift distribution. Of course, it may turn out that their distances will ultimately be measured using photometric redshifts. However, photometric redshifts will carry little credibility until they have been reliably calibrated, as we have almost no basis for understanding the stellar populations of these very faint sources.

3.5.5.1 Transformation of the Correlation Function

These very faint sources should exhibit angular clustering which is measurable on the scale of a few arc seconds, despite the dilution of the clustering amplitude due to integration along the line of sight (see, e.g., [46] for a measurement of the clustering of faint galaxies in the northern Hubble Deep Field). The correlation function, $w(\theta, S)$ for sources with flux S should, of course, be isotropic on the sky when averaged over a large enough area.

When we observe these clustered sources behind a rich cluster of galaxies, their positions and magnitudes will change due to lensing. In principle, we could detect the presence of the cluster through a reduction in the density of galaxies at a given limiting magnitude [4]. However, it is not possible to perform accurate enough photometry to make this measurement when the sources only occupy a few pixels and it turns out to be easier to detect the effect of the galaxy shear field on the galaxy locations alone. If we know the convergence and shear field of the cluster, $\kappa(\theta), \gamma(\theta)$ respectively [31], then the correlation function will become anisotropic [45] (Note that if the galaxy distribution on the sky were completely random, then there would be no effect). Put simply, the positions of galaxy triples (with separations of order a few arc seconds) can take the place of individual, brighter galaxy images. These triples should be more plentiful than galaxies which are large enough for their shapes to be measured. In terms of the angular correlation function, $w(\theta; S)$ will be transformed according to

$$w[\theta; S] \to [(1-\kappa)^2 - \gamma^2] w \left[[\theta_1^2 (1-\kappa+\gamma)^2 + \theta_2^2 (1-\kappa-\gamma)^2]^{1/2}; ((1-\kappa)^2 - \gamma^2) S \right] \tag{3.27}$$

where $\theta_{1,2}$ are components resolved along the principle axes of the local magnification tensor. Close to the critical curve, the distortions are of order unity. Typically there are thousands of faint sources behind rich clusters and many hundreds of these lie close to regions of large shear.

If we linearize (3.27), then the correlation function becomes

$$w[\theta; S] \to (1 - 2\kappa) w[(1 - \kappa - \gamma \cos 2\phi)\theta; (1 - 2\kappa)S] \tag{3.28}$$

where ϕ is the the angle between the separation vector and the principal axis of shear.

Note that κ, γ depend on the comoving angular diameter distance to the source and cluster, D_S, D_L respectively, through

$$\kappa, \gamma \propto (1 - D_L/D_S) \tag{3.29}$$

assuming a flat universe. In principle, it should be possible to invert (3.28) to obtain the distribution $dN/dD_S(D_S; S)$ for the source galaxies both by using an individual cluster and, more reliably, through combining results from different clusters at different redshifts. However, the optimal way to approach this is not yet understood. A pilot investigation on the cluster A2218 shows that $\sim 28^m$ sources with $L \sim 0.1L^*$ appear to be clustered on ~ 10 kpc scales and that their correlation function seems to exhibit the anisotropy expected from the shearing effect of the intervening cluster. There are roughly a thousand sources behind a rich cluster like A2218 which can be used for this purpose. One reason why this method gives a relatively strong signal is that the direction of the shear is independent of the source redshift. A more detailed account of this preliminary study will be presented elsewhere.

3.5.5.2 Field Galaxy Lensing Constraints on $N(z)$

Previous work using lensing by massive clusters has been placed some statistical constraints on the shape of the redshift distribution of galaxies, $N(z)$, for galaxies with $z \gtrsim 1$ (see, e.g., [24]). However, it may be that weak lensing by field galaxies could provide even stronger constraints simply because the number of sources is potentially so much larger than would be seen through a cluster lens. That is, given a sufficiently deep field in which the redshifts and luminosities of foreground lens galaxies are known, and assuming that one understands the gravitational potentials of the lenses sufficiently well, then the galaxy–galaxy lensing signal in the field should be useful for constraining the shape of the redshift distribution of the faint galaxies for which neither the redshift nor the luminosity is known.

As a preliminary example of such an investigation, we will consider the region of the Hubble Deep Field (North) [49] which has been the subject of deep redshift survey [6]. In addition to the spectroscopy of ~ 700 galaxies, extensive multi-color photometry has been obtained [19] and, hence, both the distances and restframe blue luminosities of these galaxies are "known". By a simple scaling of the halo velocity dispersions with intrinsic luminosity (i.e., a Faber–Jackson relationship), it is then straightforward to predict the shear field that these galaxies with known redshifts and luminosities would produce. Shown in Fig. 3.5 is the result obtained for a cosmology in which $\Omega_0 = 0.3$, $\Lambda_0 = 0.7$, $H_0 = 65$ km/s/Mpc, and the halos of L^* galaxies are assumed to have velocity dispersions of 155 km/s and truncation radii of $50h^{-1}$ kpc. These predictions are based upon the very simple assumption that all source galaxies lie in a single plane in redshift and the plane has been varied from $z_s = 0.5$ to $z_s = 2.0$ (i.e., Fig. 3.5 simply shows the increasing complexity of the shear field as the redshift of the sources is varied).

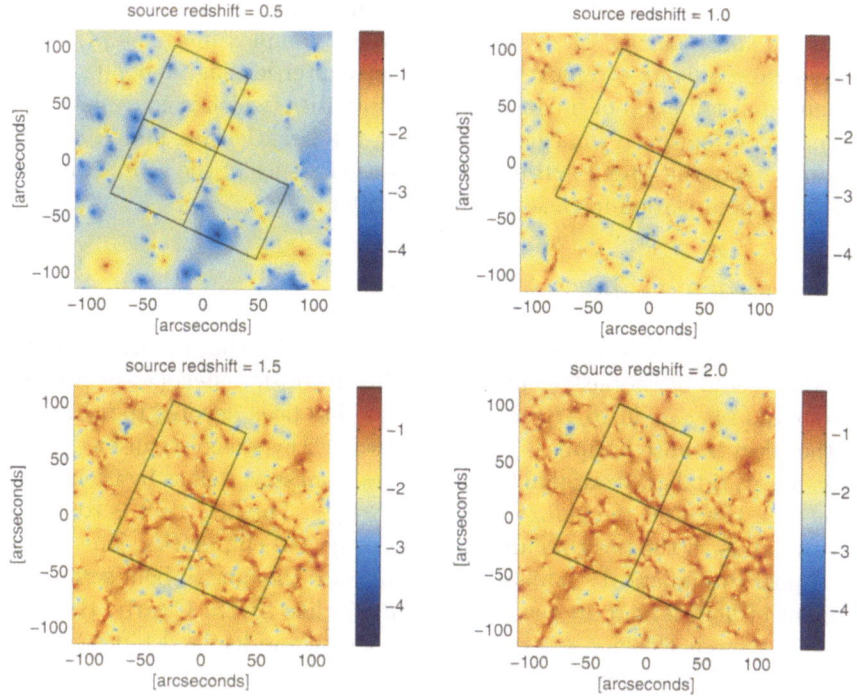

Fig. 3.5. The theoretical shear field in the region of the Hubble Deep Field (North). The grayscale indicates the logarithm of the shear. The lenses used in the calculation are galaxies in the HDF and flanking fields for which redshifts are publicly–available. In each of the panels the sources are assumed to lie in a single plane with redshift z_s, where z_s ranges from 0.5 to 2.0 (see Color Plate).

By using a maximum likelihood technique it should be possible to constrain the shape of the redshift distribution of the faint galaxies in the HDF by determining the most probable redshift for each galaxy, given its observed shape parameters (in addition to the lensing properties of the galaxies with known redshifts). That is, the observed (complex) image ellipticity is known for each of the source galaxies: $\chi = \epsilon\, e^{2i\phi}$, where $\epsilon = (a^2 - b^2)/(a^2 + b^2)$ and ϕ is the observed position angle. In the weak lensing limit the observed complex image ellipticity, χ, is just the sum of the intrinsic complex image ellipticity, χ_0, and the net distortion due to all foreground lenses. For any redshift that one might adopt for a given source, then, one can compute its *intrinsic* complex ellipticity since both the observed complex ellipticity and the image polarization for that source redshift are effectively *known*. Having done this, one can then ask the question "What is the likelihood of observing a particular source galaxy shape if the galaxy is actually located at a redshift, z?" and by maximizing a likelihood function of the form

$$\mathcal{L}(\chi,\phi,z) \equiv \frac{\partial(\chi_0,\phi_0)_{\boldsymbol{p}(z)}}{\partial(\chi,\phi)_{\boldsymbol{p}(z)}}\frac{f(\chi_0)}{2\pi} = \frac{\chi}{\chi_0}\frac{f_0(\chi_0)}{2\pi}, \tag{3.30}$$

one can arrive at a statistical constraint on the shape of the redshift distribution of the faint galaxies. Here χ_0 is the intrinsic complex image ellipticity of a source galaxy, χ is its complex image ellipticity after lensing, and $f_0(\chi_0)$ is the normalized distribution of intrinsic image ellipticities (which is equivalent to the distribution of observed image ellipticities in the weak lensing limit), ϕ and ϕ_0 are respectively the lensed and unlensed position angles of the source, and $\boldsymbol{p}(z)$ is the local polarization for a source at redshift z.

3.6   Intrinsic Galaxy Alignment

There has recently been much interest in the possibility that optical, weak lensing signals could be seriously contaminated by "intrinsic" effects associated with the sources (see, e.g., [5,16,37]). Most attention has been devoted to field lensing and the effect is quite complicated. Firstly, two nearby galaxies can be deformed by the linear, tidal gravitational field of a distant mass as well as by their mutual interaction. However, the galaxy is likely to rotate. The net effect is to create a net correlation in the position angles of the projected ellipticities. These effects also appear in numerical simulations and they are thought to be a serious contaminant of field lensing in low redshift surveys like SDSS. Clearly, the key to removing them is efficient separation of the source and lens redshifts.

When we turn to galaxy-galaxy lensing, there are analogous concerns. Here, the intrinsic effect is that the sample of distant source galaxies may be contaminated by galaxies that are interacting with the lens. Most galaxy encounters will occur with enough angular momentum to ensure that the Newtonian tidal gravitational field will stretch the galaxy preferentially in the tangential direction, especially just after perigalacticon. Furthermore, these encounters are likely to engender star formation which can make photometric redshifts misleading and make source–lens separation quite difficult. These effects will have to be modeled through numerical simulations.

3.7   Conclusions

In spite of the impressive advances in N-body, cold dark matter simulations and in the augmentation of these calculations with gas dynamical routines, the study of galaxy (and cluster) formation is essentially an empirical business. Historically, the scorecard on prediction is only average. Galaxies, clusters and quasars are observed earlier in the evolution of the universe than was generally predicted. Elliptical galaxies are more common at high redshift than expected and there is less evidence of merging than simulations show. The shapes of the potential wells also appear to be less cuspier than exhibited by numerical simulations. Of course there are mitigating circumstances such as the almost completely unexpected discovery of dark energy and there is the possibility of retrofitting these simulations with semi-analytic investigations.

We are now entering a new phase of investigation where we will be dealing with even more problematic physical processes than those that have occupied cosmologists to date – the subtleties of star formation, molecular chemistry, the inhomogeneous build up of metals, the environmental impact of AGN and minor mergers, the dynamical influence of shock-accelerated cosmic rays, subtle radiative transfer effects that limit the growth of the ionizing radiation field, and so on.

In view of these uncertainties it seems that the measured distribution of dark matter, on scales where it dominates the gas, is the primary point of contact between theory and observation. It is here that weak lensing studies will supply the most reliable information. If, as now seems likely, we can be confident about the cosmography and the form of the initial potential perturbations, then we have a framework in which to discuss early star formation and galaxy assembly.

From a theoretical viewpoint, there are really three regimes to consider. On the small scale from 3-300 kpc, say, we are mostly dealing with relaxed dark matter halos surrounding individual galaxies where gas and stars are presumably a minor contributor to the potential. By contrast, on the large scale, $\sim 10 - 300$ Mpc, the correlation function is not large and we expect little fractionation between dark and baryonic matter. We can anticipate considerable progress in measuring and interpreting statistical weak lensing on both of these scales. This should characterize quite accurately the dark matter distributions around galaxies of different morphologies, ages and in specific environments and pin down the contemporary, density fluctuation spectrum in the linear and mildly nonlinear regime. The Hubble Space Telescope will soon be equipped with the "Advanced Camera for Surveys" (ACS) which will greatly increase the quality and quantity of galaxy images suitable for weak lensing analyses. It is likely that at least a square degree of the sky will be observed to depths $I \sim 25$, which should yield roughly 200,000 sources and perhaps 60,000 lenses. The former are likely to have redshifts mostly in the range $1 < z < 2$ and the sample should be relatively free of intrinsic alignment.

It is on the intermediate or "mesoscale", $\sim 0.3 - 10$ Mpc, that the challenge will be the greatest. Here, we know that the density fluctuations are very large and quite non-linear and that the halo shapes must be quite irregular. (Galaxies only move ~ 3 Mpc at their virial speeds in ~ 10 Gyr.) Individual groups and filaments of dark matter will reflect the history of past encounters and the hierarchical assembly of dark halos. The best approach to studying this mesoscale statistically will be the direct comparison of quite sophisticated measures of the observed shear field with the results from numerical N-body simulations ([17]. (It will not be necessary to include hydrodynamical effects in these simulations.) There is much scope for experimentation in advance of the acquisition of the large datasets that are imminent.

Looking to the future, one long term goal must surely be to measure enough background sources to map the dark matter surface density around *individual* groups, and ultimately galaxies, in the manner that has already happened with clusters of galaxies. Using (cf (3.9)), and assuming that $\langle \epsilon^{-1} \rangle \sim 3$, we find that the signal to noise in a shear measurement of an individual halo is $\sim (\sigma_v/300 \text{ km s}^{-1})(n_s/10 \text{ min}^{-2})^{1/2}$, where n_s is the sky density of galaxies whose position angles can be measured reliably. This estimate is independent of the distance to the halo and the scale on which the measurement is made. It implies that useful studies of individual lens galaxies, with $S/N \sim 10$, will require source shape measurements to $i \sim 28$. This may be just outside the range of the ACS, but could be a realistic goal for the Next Generation Space Telescope. However studies of individual groups should become possible sooner and, by combining surface density measurements with velocity data and surface photometry, it may be possible to reconstruct how the constituent galaxies actually came into being. The future of weak lensing studies of dark halos looks bright.

Acknowledgments

We wish to thank Frederic Courbin, Dante Minniti, and Hernan Quintana for their hospitality and patience. Support by the US National Science Foundation under grants AST-9616968 (TGB) and AST-9900866 (RDB) is gratefully acknowledged.

References

1. M. Bartlemann & P. Schneider Phys. Rep. (2001)
2. T. G. Brainerd, R. D. Blandford & I. Smail: ApJ **466**, 623 (1996), BBS
3. T. G. Brainerd & C. O. Wright: submitted to PASP, astro-ph/0006281 (2001)
4. T. Broadhurst, A. N. Taylor, & J. Peacock: ApJ **438**, 49 (1995)
5. P. Catelan, M. Kamionkowski & R. D. Blandford: MNRAS **320**, 7 (2001)
6. J. G. Cohen, D. W. Hogg, R. D. Blandford, et al.: ApJ **538**, 29 (2000)
7. I. P. Dell'Antonio & J. A. Tyson: ApJ **473**, L17 (1996)
8. J. Dubinski & R. G. Carlberg: ApJ **378**, 496 (1991)
9. T. Ebbels: Galaxy Evolution from Gravitational Lensing Studies with the Hubble Space Telescope, PhD Thesis, University of Cambridge (1998)
10. A. Einstein: Annalen der Physik **35**, 898 (1911)
11. A. Einstein: Sitzungber. Preuss. Akad. Wissensch. erster Halbband, p. 831 (1915)
12. P. Fischer et al. (the SDSS collaboration): AJ **120**, 1198 (2000)
13. B. Geiger & P. Schneider: MNRAS **302**, 118 (1999)
14. S. Ghigna, B. Moore, F. Governato, G. Lake, T. Quinn & J. Stadel: ApJ **544**, 616 (2000)
15. R. E. Griffiths, S. Casertano, M. Im, et al.: MNRAS **282**, P1159 (1996)
16. A. Heavens, A. Refregier & C. Heymans: MNRAS **319**, 649 (2000)
17. H. Hoekstra: A Weak Lensing Study of Massive Structures. PhD Thesis, University of Groningen (2000)
18. H. Hoekstra, H. K. C. Yee & M. D. Gladders: ApJ **558**, L11 (2001)
19. D. W. Hogg, M. A. Pahre, K. L. Adelberger, et al.: ApJS **127**, 1 (2000)
20. M. J. Hudson, S. D. J. Gwyn, H. Dahle, & N. Kaiser: ApJ **503**, 531 (1998)
21. A. O. Jaunsen: Gravitational Lensing and Gamma-Ray Bursts as Cosmological Probes, PhD Thesis, University of Oslo (2000)
22. N. Kaiser: MNRAS **227**, 1 (1987)
23. N. Kaiser: ApJ **388**, 272 (1992)

24. J.-P. Kneib, R. S. Ellis, I. Smail, et al.: ApJ **471**, 643 (1996)
25. C. S. Kochanek, E. E. Falco, C. D. Impey, J. Lehár, B. A. McLeod, H.-W. Rix, C. R. Keeton, J. A. Muñoz, C. Y. Peng: ApJ **543**, 131 (2000)
26. R. Kormann, P. Schneider & M. Bartelmann: A&A **284**, 285 (1994)
27. J. Loveday, S. J. Maddox, G. Efstathiou, et al.: ApJ **442**, 457 (1995)
28. A. H. Maller, L. Simard, P. Guhathakurta, J. Hjorth, A. O. Jaunsen, R. A. Flores, & J. R. Primack: ApJ **533**, 194 (2000)
29. T. A. McKay: private communication (2000)
30. T. A. McKay, E. S. Sheldon et al.: submitted to ApJ, astro-ph/0108013 (2001)
31. Y. Mellier: ARA&A **37**, 127 (1999)
32. B. Moore *et al*: ApJ B. Moore ApJ **535**, L21 (2000)
33. P. Natarajan, J.-P. Kneib, I. Smail, & R. S. Ellis: ApJ **499**, 600 (1998)
34. P. Natarajan, J.-P. Kneib, & I. Smail: in *Gravitational Lensing: Recent Progress & Future Goals*, ASP Conf. Series 237, eds. T. G. Brainerd & C. S. Kochanek, 391 (2001)
35. P. Natarajan & A. Refregier: ApJ **538**, L113 (2000)
36. J. Navarro, C. Frenk & S. D. M. White: ApJ **490**, 493 (1997)
37. U.-L. Pen, J. Lee & U. Seljak: ApJ **543**, L107 (2000)
38. P. Sackett: in *Galaxy Dynamics*, ASP Conf. Series 182, eds. D. R. Merritt, M. Valluri, & J. A. Sellwood, 393 (1999)
39. R. Sancisi: Ap&SS **269**, 59 (1999)
40. P. Schneider & H.-W. Rix: ApJ **474**, 25 (1997)
41. I. Smail, D. W. Hogg, L. Yan, & J. G. Cohen: ApJ **449**, L10 (1995)
42. Soldner, J.: Berliner Astron. Jahrb. 1804, p. 161 (1804)
43. D. Smith, G. M. Bernstein, P. Fischer, M. Jarvis: ApJ **551**, 643 (2001)
44. J. A. Tyson, F. Valdes, J. F. Jarvis & A. P. Mills: ApJ **281**, L59 (1984)
45. L. Van Waerbeke, Y. Mellier, P. Schneider, B. Fort & G. Mathez: Astron. Astrophys. 342 15 (1997)
46. J. V. Villumsen, W. Freudling & L. N. da Costa: ApJ **481**, 578 (1997)
47. M. S. Warren, P. J. Quinn, J. K. Salmon, & W. H. Zurek: ApJ **399**, 405 (1992)
48. D. Walsh, R. F. Carswell& R. Weymann: Nature **279**, 381 (1979)
49. R. E. Williams, B. Blacker, et al.: AJ **112**, 1335 (1996)
50. G. Wilson, N. Kaiser, & G. Luppino: ApJ **555**, 572 (2001)
51. G. Wilson: private communication (2001)
52. C. O. Wright: Applications of Weak Gravitational Lensing: Constraining the Dark Matter in Clusters and Galaxies, PhD thesis, Boston University (2001)

4 Gravitational Lensing at Millimeter Wavelengths

Tommy Wiklind[1,*] and Danielle Alloin[2]

[1] Dept. of Astronomy & Astrophysics, Onsala Space Observatory,
SE-43992 Onsala, Sweden
[2] European Southern Observatory, Casilla 19001 Santiago 19, Chile

Abstract. The study of gas and dust at high redshift gives an unbiased view of star formation in obscured objects as well as the chemical evolution history of galaxies. With today's millimeter and submillimeter instruments observers use gravitational lensing mostly as a tool to boost the sensitivity when observing distant objects. This is evident through the dominance of gravitationally lensed objects among those detected in CO rotational lines at $z > 1$. It is also evident in the use of lensing magnification by galaxy clusters in order to reach faint submm/mm continuum sources. There are, however, a few cases where millimeter lines have been directly involved in understanding lensing configurations. Future mm/submm instruments, such as the ALMA interferometer, will have both the sensitivity and the angular resolution to allow detailed observations of gravitational lenses. The almost constant sensitivity to dust emission over the redshift range $z \approx 1 - 10$ means that the likelihood for strong lensing of dust continuum sources is much higher than for optically selected sources. A large number of new strong lenses are therefore likely to be discovered with ALMA, allowing a direct assessment of cosmological parameters through lens statistics. Combined with an angular resolution $< 0\farcs1$, ALMA will also be efficient for probing the gravitational potential of galaxy clusters, where we will be able to study both the sources and the lenses themselves, free of obscuration and extinction corrections, derive rotation curves for the lenses, their orientation and, thus, greatly constrain lens models.

4.1 Introduction

Rapid progress in the development of millimeter astronomical facilities, such as the increase of antennae sizes and/or of the number of array elements, or as the continuing improvement in the sensitivity of detectors, have now made it possible to explore the high redshift universe in this window and therefore to exploit the potentialities of gravitational lensing effects.

Why is it so important to explore this wavelength domain? In one short statement: the presence of cold dust and of molecular material can be traced in this window and both components witness the formation of heavy elements. If they are detected in galaxies at high redshifts, they allow us to probe star formation in the early universe. They reveal as well processes related to the startup of Active Galactic Nuclei (AGN) and signal the presence of massive black-holes.

---

* Now affiliated at: STScI ESA Space Telescope Division, 3700 San Martin Dr., Baltimore, MD 21218, USA

Large amounts of molecular gas are encountered in the close environment of the central engine in AGN. This material is often regarded as the fuel which allows to activate the AGN. An evolutionary scenario would then connect IR-luminous galaxies rich in molecular material and with intense star formation to the formation/feeding of massive black holes.

Several fundamental questions are therefore underlying the search for dust and molecular gas at high redshift: the redshift of galaxy formation? the chemical evolution of the universe with time? the evolution of the dust content in the universe at early ages? the epoch and the scenario of the formation of massive black holes? the startup and evolution of AGN activity? Do we have already some clues to answer these questions? The most powerful AGNs, quasars, are now detected up to redshift around 6.5, and galaxies up to redshift 6. So we know that in this redshift range the universe already hosted galaxies and massive black holes and that its metal content was substantial since the spectra of high redshift AGNs are very similar to those of low redshift AGNs. Yet, only a few objects are known at these high redshifts and this may provide a biased view. It is therefore mandatory to enlarge the sample and of course, the goal is also to push the redshift limit.

Pushing the redshift limit also means that we are investigating sources with lower and lower flux density. This can be achieved through technical improvements, using larger collectors and better detectors. The ALMA (Atacama Large Millimeter Array) project is showing the way. Another manner is to take advantage of the effects induced by gravitational lensing (for a review of its theoretical basis, see the comprehensive book by Schneider et al. [126]). Firstly, image magnification allows us to detect more distant sources of cold dust and molecular gas of a given intrinsic luminosity, or to detect at a given redshift sources of fainter intrinsic luminosity. The latter in particular is important for good determinations of luminosity functions. Secondly, differential magnification effects can be used as an elegant tool to probe the size of molecular and dusty structures in the lensed source, as long as the lensing system provides the appropriate geometry. In this case, it is imperative to have an excellent model of the lensing system, as any structural information about the source itself for example is recovered by tracing the image back through the lensing system. Both aspects will be discussed at length in this paper. In some cases molecular absorption lines allow us to obtain information about the lensing galaxy itself.

Apart from hydrogen and helium, carbon and oxygen are the heavy elements with highest abundance in the universe. Therefore, the CO molecule is the most suitable candidate for the detection of molecular gas in emission at high redshift. The CO molecules can be detected directly through their thermal line emission in the source or as silhouetted absorbers along the line of sight to a background source. The latter may occur for example for the lensing galaxy. Several other molecules have been detected at high redshift, HCO^+, HCN, H_2CO..., while dust is detected essentially through its thermal emission.

We provide in Sect. 4.2 an overview of the CO line emission and of the high redshift CO sources detected so far. The role played by gravitational lensing in studying CO sources at high redshift is highlighted. In Sect. 4.3 we review molecular absorption and the importance of such measurements to investigate the properties of the lensing galaxies. Sect. 4.4 introduces the dust continuum emission and the use of differential magnification effects which can be made to probe the dust content of the lensed objects. Three cases particularly well studied are discussed in detail in Sect. 4.5. In Sect. 4.6 existing lens models for PKS 1830-211 are reviewed and a new one introduced. Finally, in Sect. 4.7 the future of this type of investigations is presented in the perspective of new instrumental developments in general and of ALMA in particular.

4.2 Molecular Emission

The goal of this section is primarily to highlight the benefits of exploiting gravitational lensing effects in the millimeter range, that is in CO line emission. Therefore, after some brief comments on the pioneering observations of low redshift sources, we shall concentrate on the results obtained on high redshift sources. This section deals essentially with the CO line emission, while the following sections will discuss molecular absorption lines and dust thermal emission.

Detection and measurements of the ^{12}CO rotational transitions in Galactic and extragalactic sources have had a great impact on the development of astrochemistry. The J=1-0 CO transition is excited by collisions with H_2 molecules, even in clouds at low kinetic temperature. The low-J transitions are in general optically thick, the opacity being determined observationally by the relative intensity of the corresponding ^{13}CO transition. On the contrary, the high-J transitions are optically thin. Therefore, it is quite interesting to perform multi-transition studies to ascertain in a more secure fashion the physical conditions, temperature and density, in the emitting molecular material. This is particularly true in the case of very dense molecular material exposed to intense radiation fields, like in the environment of an AGN or in powerful star forming regions. The conversion factor which is used to derive the total mass of molecular material from the observed L_{CO}, is also highly dependent on the physical conditions in the molecular material. It has been determined to be 4.6 $M_\odot$ (K km s^{-1} pc^2)$^{-1}$ in standard Milky Way clouds [138]. Recently it has been shown that lower values of the conversion factor (by up to a factor 10) should be used in the case of dense and warm material as usually encountered in a molecular torus around an AGN [41,13].

4.2.1 Low- and Intermediate Redshift Galaxies

Following far-infrared (FIR) observations by IRAS, an important population of IR-luminous (dust-rich) galaxies was found. This was the starting point for investigating as well their molecular content, using the millimeter facilities available in the late 80's. A number of IR-luminous galaxies and AGNs were detected, mostly in the CO(1-0) transition and the field develop quickly. Regarding low redshift sources, let us briefly mention the detection of AGNs such as Mrk 231

[123], Mrk 1014 [124], I Zw 1 [9] and of some low redshift quasars [125,1] or radio galaxies [114,101,100]. At moderate redshift, CO was detected in the radio galaxy 3C48 ($z = 0.369$) [127]. This search is continuing through e.g. the Caltech CO high and low redshift radio galaxy survey [49].

On the side of CO sources at high redshift ($z>2$), the first object detected was IRAS 10214+4724 [23,136]. All along the 90's, millimeter dishes and interferometer arrays in service were pushed to their limits in searching for other candidates, selected for example upon the strength of their submillimeter flux. A large amount of observing time was dedicated to such programs at the OVRO, BIMA, Nobeyama and IRAM facilities. At face value, the success rate in detecting high redshift CO sources has been modest. One reason for this is the uncertainty in the precise redshift of the emitting molecular gas under search, while the backends of the instruments are narrow in comparison to the redshift range to be explored. Another reason is of course the limited sensitivity of existing instruments. Only the most luminous and most gas-rich systems can be detected. The situation should improve with new facilities such as ALMA (See Sect. 4.7.1). Sources detected so far are detailed below, in order of increasing redshift.

Table 4.1. Galaxies at z>1 with molecular emission (January 2002)

Name	z	$M_{H_2}/M_\odot$	$M_d/M_\odot$	$L_{FIR}/L_\odot$	Grav. lens	Ref.
BR1202−0725	4.69	6×10^{10}	2×10^8	$\sim 1 \times 10^{12}$	?	[103],[105]
BR0952−0115	4.43	3×10^9	3×10^7	$\sim 1 \times 10^{12}$	YES	[61]
BRI1335−0414	4.41	1×10^{11}	continuum det.	—	NO?	[60]
PSS2322+1944	4.12	3×10^{11}	1×10^9	—	?	[38]
APM 08279+5255	3.91	2×10^9	1×10^7	$\sim 8 \times 10^{13}$	YES	[43]
4C60.07	3.79	8×10^{10}	2×10^8	$\sim 2 \times 10^{13}$	NO	[107]
6C1909+722	3.53	4×10^{10}	2×10^8	$\sim 2 \times 10^{13}$	NO	[107]
MG0751+2716	3.20	8×10^{10}	—	—	YES	[10]
SMM02399−0136	2.81	8×10^{10}	continuum det.	$\sim 1 \times 10^{13}$	YES	[53]
MG0414+0534	2.64	5×10^{10}	continuum det.	—	YES	[11]
SMM14011+0252	2.56	5×10^{10}	continuum det.	$\sim 3 \times 10^{12}$	YES	[54]
H1413+117	2.56	2×10^9	1×10^8	$\sim 2 \times 10^{12}$	YES	[12]
53W002	2.39	1×10^{10}	weak continuum	—	NO	[129]
F10214+4724	2.28	2×10^{10}	9×10^8	$\sim 7 \times 10^{12}$	YES	[23]
HR 10	1.44	7×10^{10}	2×10^8	$\sim 9 \times 10^{11}$	NO	[4]

Masses and fluxes corrected for gravitational magnification (approx.)

4.2.2 High Redshift Galaxies

The first source discovered, IRAS 10214+4724, at $z = 2.285$, has been detected in CO(3-2), CO(4-3) and CO(6-5) [23,136]. A report on the detection of CO(1-0) [145] remains to be confirmed. The source is a gravitationally lensed ultraluminous IR-galaxy [22,40]. The magnification factor is found to be around 10 for the CO source which has an intrinsic radius of 400 pc. Conversely, the far-IR emission detected in this object is magnified 13 times and arises from a source with radius 250 pc, while the mid-IR is magnified 50 times and arises from a source with radius 40 pc. After correcting for magnification, and using a conversion factor L'_{CO} to $M(H_2)$ of 4 $M_\odot$ (K km s^{-1} pc^2)$^{-1}$ [119], the molecular gas mass is found to be 2×10^{10} $M_\odot$, in agreement with the estimated dynamical mass 3×10^{10} $M_\odot$. As noted above however, such a value for the conversion factor, obtained from CO(1-0) observations of Galactic molecular clouds, might not be applicable in the case of warmer and denser molecular material. A value for the conversion factor which is 5 times lower than the standard one has been found in a study of extreme starbursts in IR-luminous galaxies [41]. Hence, the mass value quoted above should be regarded as an upper limit to the mass of the molecular gas in IRAS 10214+472. Still pending is the question of the CO line emission share between a hidden AGN and a starburst in the 400 pc region surrounding the AGN. We notice also that the large extension (3 to 12 kpc) in CO emission reported by [128] has not been confirmed by the IRAM interferometer data [40].

One interesting case is that of the radio galaxy 53W002 at $z = 2.394$, located at the center of a group of $\sim$20 Lyman-α emitters. A possible detection of the CO(1-0) line at Nobeyama was reported in [163], although not yet confirmed by others. The first detection in CO(3-2) by OVRO [129] suggested a large extension (30 kpc) and the existence of a velocity gradient. None of these features has been confirmed by an IRAM interferometer data set with higher signal to noise ratio [2]. From an astrometric analysis it is found that the 8.4 GHz and CO source are coincident, at a location consistent with that of the optical/UV continuum source. The most likely origin of the molecular emission is therefore from the close environment of the AGN. One should notice that 53W002 is definitely not a gravitationally lensed source. Using a conversion factor L'_{CO} to $M(H_2)$ in the range 0.4 - 0.8 $M_\odot$ (K km s^{-1} pc^2)$^{-1}$, more appropriate for dense and warm molecular gas around an AGN [13], the resultant molecular gas mass is found to be in the range $(0.6 - 1.0) \times 10^{10}$ $M_\odot$.

The Cloverleaf, H1413+117 is a well known gravitationally lensed Broad Absorption Line (BAL) quasar at $z = 2.558$ [96]. Its CO(3-2) transition was first observed with the IRAM 30m dish [12] and then with BIMA [160]. Later, the CO(4-3), CO(5-4) and CO(7-6) transitions have been detected, together with HCN(4-3) and a fine-structure line of CI. From a detailed analysis of these transitions, the molecular gas was found to be warm and dense [13] with a low conversion factor of 0.4 $M_\odot$ (K km s^{-1} pc^2)$^{-1}$. High resolution maps in the CO(7-6) transition were obtained with OVRO [164] and with the IRAM interferometer [3,77]. The IRAM map has the best resolution and signal to noise ratio. Comparing with the HST images and exploiting differential gravitational effects, the CO(7-6) map allowed a derivation of both the size and the kinematics of the molecular/dusty torus around the quasar central engine [77] (for further details see Sect. 4.5.2). After correcting for the amplification factor (30 according to the model used for the lensing system), and using the mean conversion factor 0.6 $M_\odot$ (K km s^{-1} pc^2)$^{-1}$ (derived by Barvainis et al. [13]), the derived mass of molecular gas $M(H_2)$ is 2×10^9 $M_\odot$, in agreement with the dynamical mass of 8×10^8 $M_\odot$.

The source SMM 14011+0252 was observed with OVRO in CO(3-2) at $z = 2.565$ [54], following its discovery as a strong submillimeter source detected in the course a survey of rich lensing clusters [134]. Correcting for an amplification factor of 2.75 and using a conversion factor of 4 $M_\odot$ (K km s^{-1} pc^2)$^{-1}$, the mass of molecular gas turns out to be 5×10^{10} $M_\odot$, while the dynamical mass is found to be larger than 1.5×10^{10} $M_\odot$. The CO emission is extended on scales of $\sim$10 kpc and associated with likewise extended radio continuum emission [73]. Optical and near-infrared (NIR) imaging shows two objects (designated J1 and J2), separated by $2''.1$ [72]. Both the molecular gas and the radio continuum, however, have their strongest emission $\sim 1''$ north of the J1/J2 components and are extended between J1 and J2. This suggests that the optical/NIR emission comes from two 'windows' in the obscuring molecular gas and dust and that J1/J2 represent emission from a coherent large galaxy. The extended nature of the radio continuum, the lack of X-ray emission [50] and the lack of optical broad emission lines (Wiklind et al. 2002 in prep) suggest that only star formation powers the large FIR luminosity. Assuming a Salpeter initial mass function and correcting for the gravitational magnification, the FIR luminosity indicates a star formation rate exceeding 10^3 $M_\odot$ yr^{-1}.

The gravitationally lensed quasar MG 0414+0534 was observed with the IRAM interferometer in the CO(3-2) transition at $z = 2.639$ [11]. The lensed nature of this system is known from a 5 GHz map [63]. It displays four quasar-spots separated at most by $\sim 2''$. The beam of the IRAM data ($2''.0 \times 0''.9$) does not allow to separate the components in the CO(3-2) velocity-integrated map. However, by fitting the UV data directly, it has been possible to resolve the combined A components (A1+A2) from component B and to get separate CO(3-2) spectra. The relative strength A:B in the 5 GHz radio continuum is 5:1. The millimeter continuum rather shows a ratio 7:1 and differences are seen between the A and B CO(3-2) spectra, suggesting that differential magnification effects may be at work. The magnification factor is unknown: hence only an upper limit can be derived for the molecular gas mass. Assuming in addition a conservative conversion factor of 4 $M_\odot$ (K km s^{-1} pc^2)$^{-1}$, the upper limit found for $M(H_2)$ is 2.2×10^{11} $M_\odot$. This figure is below the upper limit derived for the dynamical mass, 9×10^{11} $M_\odot$.

Another source detected first through submillimeter observations is SMM 02399-0136 [71]. It is known to be gravitationally amplified by a foreground cluster of galaxies, the amplification factor being 2.5. This source has been detected with OVRO in the CO(3-2) transition at $z = 2.808$ [53]. The mass of molecular gas deduced in this object, correcting for a 2.5 amplification factor and using the conversion factor applicable to Galactic clouds, 4 $M_\odot$ (K km s^{-1} pc^2)$^{-1}$, is 8×10^{10} $M_\odot$. From the upper limit on the apparent size of the CO emitting source (5″) and the width of the CO line (710 km s^{-1}), an upper limit to the mass of molecular gas of 1.5×10^{10} $M_\odot$ can be derived [53]. The SCUBA results indicate that SMM 02399-0136 is an IR-hyperluminous galaxy. On the other hand, optical data show clearly than it hosts, as well, a dust-enshrouded AGN [71]. A precise share of CO emission between the two components remains to be investigated.

In the course of a systematic CO emission survey of gravitationally lensed sources with the IRAM interferometer [10], the source MG 0751+2716 has been detected in the CO(4-3) transition at $z = 3.200$. This source was first discovered to be a gravitationally lensed quasar, from VLA maps [86]. It shows four quasar-spots with maximum separation of 0″9. The lensing galaxy, which provides the image geometrical configuration, is part of a group of galaxies adding another shear to the lens-system [144]. The lensing system remains to be modeled in detail. Therefore, the amplification factor is not known. However, given the observed strength of the CO line emission it should be large. Assuming that the CO emission in this source is mostly from the close environment of the AGN, we consider a conversion factor in the range 0.4 to 1 $M_\odot$ (K km s^{-1} pc^2)$^{-1}$. The corresponding upper limit (no correction applied for the unknown amplification factor) for the mass of molecular gas is in the range 8×10^{10} to 2×10^{11} $M_\odot$ [10].

The distant powerful radio galaxy 6C 1909+722 has been detected in the CO(4-3) line at $z = 3.53$, with the IRAM interferometer, and in dust submillimeter emission using SCUBA [107]. It is unlikely to be a gravitationally lensed object. Hence, the derived mass of molecular material is quite large, even assuming a conservative value for the conversion factor (about one fifth the value derived from Galactic molecular clouds). It is found to be in the range $(0.5 - 1.0) \times 10^{11}$ $M_\odot$.

Another possibly unlensed powerful radio galaxy 4C 60.07, has been detected in the CO(4-3) line emission at $z = 3.79$ at IRAM, and in dust thermal emission at submillimeter wavelengths with SCUBA and at millimeter wavelengths at IRAM [107]. Remarkably, the CO line emission extends over 30 kpc and breaks into two components: one corresponding to the AGN (radio core) and a second one which seems to be related to a major episode of star formation. This state of merging is speculatively interpreted as the formative stage of an elliptical host around the residing AGN. Again, the molecular mass is found to be quite large, around 10^{11} $M_\odot$.

The gravitationally lensed BAL quasar, APM 08279+5255, has been detected in the CO(4-3) and CO(9-8) transitions at $z = 3.911$, with the IRAM interferometer [43]. The CO line ratio points towards warm and dense molecular gas. Thermal emission from the dust component is also measured. Both the molecular and dust luminosities appear to be very high. Gravitational amplification is therefore suspected and has subsequently been confirmed through the detection of three optical/NIR components [85] [47] (see also Sect. 4.5.1 and

Fig. 4.13). The magnification factors for the molecular gas and dust where estimated [43]. After correcting for these factors, the dust mass is found to be in the range $(1 - 7) \times 10^7$ M$_\odot$, and the molecular gas mass in the range $(1 - 6) \times 10^9$ M$_\odot$. In this interpretation, the molecular/dusty component is in the form of a nuclear disk with radius 90-270 pc orbiting the central engine of the BAL quasar. This source looks therefore quite similar to the Cloverleaf. Recently, however, extended low-excitation CO emission (the J=1-0 and J=2-1 transitions) have been detected using the VLA [106]. This extended emission is likely to be associated with a cooler molecular component than the CO(9-8) emission.

Finally, let's discuss the four sources detected so far at z larger than 4:

PSS 2322+1944. The radio quiet quasar PSS 2322+1944 has recently been detected in the CO(5-4) and CO(4-3) transitions at a redshift of $z = 4.12$ with the IRAM interferometer [38]. The velocity-integrated CO line fluxes are 3.74 ± 0.56 and 4.24 ± 0.33 Jy km s^{-1}, with a linewidth ≈ 330 km s^{-1}. The 1.35 mm (250μm restwavelength) dust continuum flux density is 7.5 mJy, in agreement with previous measurements at 1.25 mm at the IRAM 30m telescope [104], and corresponds to a dust mass of $\approx 10^9$ M$_\odot$. With the present angular resolution of the observations, no evidence for extended emission has been found yet. The implied gas mass is estimated to be $\approx 3 \times 10^{11}$ M$_\odot$, using a conversion factor of 4.6 M$_\odot$ K km s^{-1} pc^2. The properties of PSS 2322+1944 are described in detail in [38].

BRI 1335-0415. BRI 1335-0415 was detected in the CO(5-4) transition with the IRAM interferometer at $z = 4.407$ [60]. The source does not exhibit a noticeable extension neither in the 1.35mm continuum nor in the CO line emission. In addition, there is no obvious sign of gravitational lensing on the line-of-sight to this source. The authors have derived a very large mass of molecular gas, close to 10^{11} M$_\odot$. Even with a conversion factor 3 times smaller, more appropriate for this type of object, the mass remains a few 10^{10} M$_\odot$.

BR 0952-0115. The gravitationally lensed radio quiet quasar BR 0952-0115 has been detected in CO(5-4) with IRAM facilities at $z = 4.43$ [61]. A tentative estimate of the mass of molecular material M(H$_2$) is 3×10^9 M$_\odot$. Note however that a more precise model of the lens has still to be worked out.

BR 1202-01215. BR 1202-01215 is the most distant source detected in CO. It has been reported in the CO(5-4) transition observed with the Nobeyama array [103], and in the CO(4-3), CO(5-4) and CO(7-6) lines observed with IRAM facilities [105]. The CO maps show two separate sources on the sky: one is coincident with the optical quasar (for this source CO(5-4) provides $z = 4.695$), while the other is located 4" to the North-West, where no optical counterpart is found (for this source CO(5-4) provides $z = 4.692$). At this redshift the $4''$ extension corresponds to a de-projected distance of 12-30 kpc. It is uncertain whether this object is gravitationally lensed or not. Is the North-West source a second image of the quasar? There are hints that this might be the case as a strong gravitational shear has been measured in the field (Fort and D'Odorico, private communication). Else, each of two separate sources ought to have its own heating source, AGN or starburst. Assuming no gravitational boost and using a conversion factor of 4 $M_\odot$ (K km s^{-1} pc^2)$^{-1}$, the mass of molecular gas is quoted to be 6×10^{10} $M_\odot$.

A summary of the sources properties is provided in Table 4.1: redshift, mass of molecular gas, mass of dust, total FIR luminosity and the status of the gravitationally lensed nature of the object.

This quick compilation of high redshift CO sources has prompted a number of key-issues:

(i) A major difficulty in detecting high redshift CO sources is the lack of precision in our guess for the redshift of the molecular gas. The instantaneous frequency coverage of current backends requires that the redshift is known a-priori with a precision of a few percent. Why is this condition hard to fulfill? The molecular gas emission in distant objects can arise from the close environment of an AGN. Yet, published redshifts for distant AGN are mostly measured from emission lines of highly ionized species which can be strongly affected by winds. Indeed velocity offsets of up to 2500 km s^{-1} have been observed between the CO lines and the blue-shifted CIV line for example (e.g. [43]). Hopefully, this limitation will be overcome with the next generation of backends.

(ii) An uncertain part in the interpretation of the observed CO line intensities lies with the physical state of the molecular gas and the conversion factor L'_{CO} to M(H$_2$) to be applied in the case of high redshift sources. When several CO transitions are observed (such as for IRAS10214+4724, H1413+117, APM 08279+5255 and BR 1202-0725), the physical conditions of the molecular gas can be analyzed, pointing towards warm (T$\sim$100 K), dense (a few 10^3 cm^{-3}) and moderately optically thick material. Such conditions could very well characterize molecular gas in the proximity of an AGN. Conversely, the conditions of the molecular gas in an extended starburst may be more similar to those encountered in Galactic molecular clouds. The conversion factor depends on the physical conditions of the molecular gas. It ranges from a value of 4 $M_\odot$ (K km s^{-1} pc^2)$^{-1}$ in Galactic clouds [119], to 1 $M_\odot$ (K km s^{-1} pc^2)$^{-1}$ in IR-ultraluminous galaxies [41] and possibly 0.4 $M_\odot$ (K km s^{-1} pc^2)$^{-1}$ in the surroundings of an AGN [13]. Therefore, it would be important to have some clues about the share AGN/starburst in the heating mechanism for high redshift CO sources. In that respect, the CO line width and the compactness of the source may bring some pieces of information.

(iii) Finally, the outmost efforts should be made to find out whether a source is gravitationally lensed or not, before the claim for the presence of a huge amount of molecular gas (10^{11} $M_\odot$) can be taken as a starting point for modeling. Weak shear from an intervening galaxy cluster, like in the cases of SMM 14011+0252 and SMM 02399-0136, induces a mild magnification factor in the range 2-3. Strong shear (possibly combined with weak shear), induces magnification factors of up to 30! This would decrease by one order of magnitude the mass of molecular gas derived. If, at the same time, the applicable conversion factor is on the low side of its possible values range, a reduction of the actual molecular gas mass by another order of magnitude would apply.

In conclusion, it is important to search for other high redshift CO sources and, at the same time, to investigate carefully the nature of their environment and line of sight, and the physical conditions in their molecular gas.

4.3   Molecular Absorption Lines

Another method to study molecular gas at high redshift is to observe molecular rotational lines in absorption rather than emission. Whereas emission is biased in favor of warm and dense molecular gas, tracing regions of active high mass star formation, molecular absorption lines trace excitationally cold gas. This is important since a large part of the molecular gas mass may reside in regions far away from massive star formation and therefore remain largely unobserved in emission.

Molecular absorption occurs whenever the line of sight to a background quasar passes through a sufficiently dense molecular cloud. In contrast to optical absorption lines seen towards most high redshift QSOs, molecular absorption is invariably associated with galaxies, either in the host galaxy of the continuum source or along the line of sight. In nearby galaxies molecular gas is strongly concentrated to the central regions, making the likelihood for absorption largest whenever the line of sight passes close to the center of an intervening galaxy. This, of course, means that molecular absorption in intervening galaxies is likely to be associated with gravitational lensing, and vice versa. Indeed, the only known systems of intervening absorption (B 0218+357 and PKS 1830-211) are gravitationally lensed and the absorption probes molecular gas in the lensing galaxy. Molecular absorption lines can thus be used to study the neutral and dense interstellar medium in lenses. At the present the sample of lens galaxies probed by molecular absorption lines is limited, but with the advent of a new sensitive interferometer instrument like ALMA, the number of potential candidate systems will increase substantially and make it possible to probe the molecular interstellar medium in the lens galaxies in some detail. Moreover, the molecular absorption lines provide unique kinematical information which is valuable when constructing a model of the lensed system.

4.3.1   Detectability

As mentioned above, molecular absorption traces a different gas component then emission lines. For optically thin *emission* the integrated signal I_{CO} is

$$I_{CO} = \int T_a \, dv \; \propto \; N_{tot} \, T^{-1} \, e^{-E_u/kT} \left(e^{h\nu/kT} - 1 \right) \left[J(T) - J(T_{bg}) \right] \;,$$

where N_{tot} is the total column density of a given molecular species, E_u the upper energy level of a transition with $\Delta E = h\nu$, T_{bg} is the local temperature of the Cosmic Microwave Background Radiation (CMBR) and $J(T) = (h\nu/k)(e^{h\nu/kT} -$

$1)^{-1}$. When $T \to T_{bg}$ all the molecules reside in the ground rotational state $J = 0$ and the signal disappears. For molecular *absorption* the observable is the velocity integrated opacity I_{τ_ν}:

$$I_{\tau_\nu} = \int \tau_\nu \, dv \;\propto\; N_{tot} \, T^{-1} \, \mu_0^2 \, e^{-E_l/kT} \left(1 - e^{-h\nu/kT} \right) \;,$$

where N_{tot} is again the total column density of a given molecular species, while E_l is now the lower energy level and μ_0 is the permanent dipole moment of the molecule. For the ground transition[1], $E_l = 0$, $I_{\tau_\nu} \propto N_{tot} \, T^{-1} \, \mu_0^2 \, (1 - e^{-h\nu/kT}) \approx (h\nu/k) \, N_{tot} \, \mu_0^2 \, T^{-2}$. In contrast to emission lines, the observed integrated opacity increases as the temperature T decreases.

Molecules are generally excited through collisions with molecular hydrogen H_2. The excitation temperature, T_x, therefore depends strongly on the H_2 density. The collisional excitation is balanced by radiative decay and a steady-state situation with $T_x = T_k$ requires a certain critical H_2 density. For CO, which has a small permanent dipole moment μ_0, the critical density is rather low, 4×10^4 cm^{-3}, while molecules with higher dipole moments require higher densities. For instance, for HCO$^+$ which has a dipole moment more than 30 times larger than that of CO, the critical density is 2×10^7 cm^{-3}.

The strong dependence of the opacity on the permanent dipole moment means that absorption preferentially probes low excitation gas, i.e. a cold and/or diffuse molecular gas component. If multiple gas components are present in the line of sight, with equal column densities but characterized by different excitation temperatures, absorption will be most sensitive to the gas component with the lowest temperature. The dependence of the opacity on the permanent dipole moments also means that molecules much less abundant than CO can be as easily detectable. For instance, HCO$^+$ has an abundance which is of the order 5×10^{-4} that of CO, yet it is as easy, or easier, to detect in absorption as CO. This is illustrated in Fig. 4.1, where the observed opacity of the CO(1-0) and HCO$^+$(2-1) transitions at $z = 0.25$ are compared. In this particular case, the HCO$^+$ line has a higher opacity than the CO line.

4.3.2   Observables

Analysis of the molecular absorption lines gives important information about both the physical and chemical properties of the interstellar medium. This can have implications for identifying the type of galaxy causing the absorption and, in some cases, help to identify the morphological type of lenses. In this section a short description of the analysis that can be done is presented. A more detailed description can be found in the references given in the text.

Optical Depth. The observed continuum temperature, T_c, away from an absorption line can be expressed as $T_c = f_s J(T_b)$, where f_s is the beam filling factor of the region emitting continuum radiation, T_b is the brightness temperature of

---

[1] This expression is strictly speaking only true for linear molecules.

the background source and $J(T) = (h\nu/k)/[1 - \exp(-h\nu/kT)]$ (e.g. [153]). The spatial extent of the region emitting continuum radiation at millimeter wavelengths is unknown but is certain to be smaller than at longer wavelengths. The BL Lac 3C446 has been observed with mm-VLBI and has a size $< 30\,\mu$arcseconds [88]. Since the angular size of a single dish telescope beam at millimeter wavelengths is typically $10'' - 25''$, the brightness temperature of the background source, T_b, is at least $10^9 \times T_c$. This means that the local excitation temperature of the molecular gas is of no significance when deriving the opacity. The excitation does enter, however, when deriving column densities.

Excitation Temperature and Column Density. The excitation temperature, T_x, relates the relative population of two energy levels of a molecule as: $\frac{n_2}{n_1} = \frac{g_2}{g_1} e^{-h\nu_{21}/kT_x}$, where g_i is the statistical weight for level i and $h\nu_{21}$ is the energy difference between two rotational levels. In order to derive T_x we must link the fractional population in level i to the total abundance. This is done by invoking the weak LTE-approximation[2]. We can then use the partition function $Q(T_x) = \sum_{J=0}^{\infty} g_J e^{-E_J/kT_x}$ to express the total column density, N_{tot}, as

$$N_{tot} = \frac{8\pi}{c^3} \frac{\nu^3}{g_J A_{J,J+1}} f(T_x) \int \tau_\nu dV \ ,$$

$$f(T_x) = \frac{Q(T_x)e^{E_J/kT_x}}{1 - e^{-h\nu/kT_x}} \ ,$$

(4.1)

where $\int \tau_\nu dV$ is the observed optical depth integrated over the line for a given transition, $g_J = 2J + 3$ for a transition $J \to J + 1$, and E_J is the energy of the rotational level J. By taking the ratio of two observed transitions from the same molecule, the excitation temperature can be derived. The strong frequency dependence of the column density in (4.1) is only apparent since the Einstein coefficient, $A_{J,J+1}$, is proportional to ν^3.

4.3.3 Known Molecular Absorption Line Systems

There are four known molecular absorption line systems at high redshift: z=0.25-0.89. These are listed in Table 4.2 together with data for the low redshift absorption system seen toward the radio core of Centaurus A. For the high redshift systems, a total of 18 different molecules have been detected, in 32 different transitions. This includes several isotopic species: $C^{13}O$, $C^{18}O$, $H^{13}CO$, $H^{13}CN$ and $HC^{18}O^+$. As can be seen from Table 4.2, the inferred H_2 column densities varies by $\sim 10^3$. The isotopic species are only detectable towards the systems with the highest column densities: B 0218+357 and PKS 1830-211, which are also the systems where the absorption originates in lensing galaxies. The large dispersion

---

[2] In the weak LTE-approximation $T_x \approx T_{rot}$, but the rotational temperature T_{rot} is not necessarily equal to the kinetic temperature and can also be different for different molecular species.

Table 4.2. Properties of molecular absorption line systems.

Source	$z_a^{(a)}$	$z_e^{(b)}$	N_{CO} cm^{-2}	N_{H_2} cm^{-2}	$N_{HI}^{(c)}$ cm^{-2}	$A_V'^{(d)}$	N_{HI}/N_{H_2}
Cen A	0.00184	0.0018	1.0×10^{16}	2.0×10^{20}	1×10^{20}	50	0.5
PKS 1413+357	0.24671	0.247	2.3×10^{16}	4.6×10^{20}	1.3×10^{21}	2.0	2.8
B3 1504+377A	0.67335	0.673	6.0×10^{16}	1.2×10^{21}	2.4×10^{21}	5.0	2.0
B3 1504+377B	0.67150	0.673	2.6×10^{16}	5.2×10^{20}	$< 7 \times 10^{20}$	<2	<1.4
B 0218+357	0.68466	0.94	2.0×10^{19}	4.0×10^{23}	4.0×10^{20}	850	1×10^{-3}
PKS 1830–211A	0.88582	2.507	2.0×10^{18}	4.0×10^{22}	5.0×10^{20}	100	1×10^{-2}
PKS 1830–211B	0.88489	2.507	$1.0 \times 10^{16\ (e)}$	2.0×10^{20}	1.0×10^{21}	1.8	5.0
PKS 1830–211C	0.19267	2.507	$< 6 \times 10^{15}$	$< 1 \times 10^{20}$	2.5×10^{20}	<0.2	>2.5

(a) Redshift of absorption line.
(b) Redshift of background source.
(c) 21cm HI data taken from [25] [26] [27] [28]. A spin-temperature of 100 K and a area covering factor of 1 was assumed.
(d) Extinction corrected for redshift using a Galactic extinction law.
(e) Estimated from the HCO$^+$ column density of 1.3×10^{13} cm^{-2}.

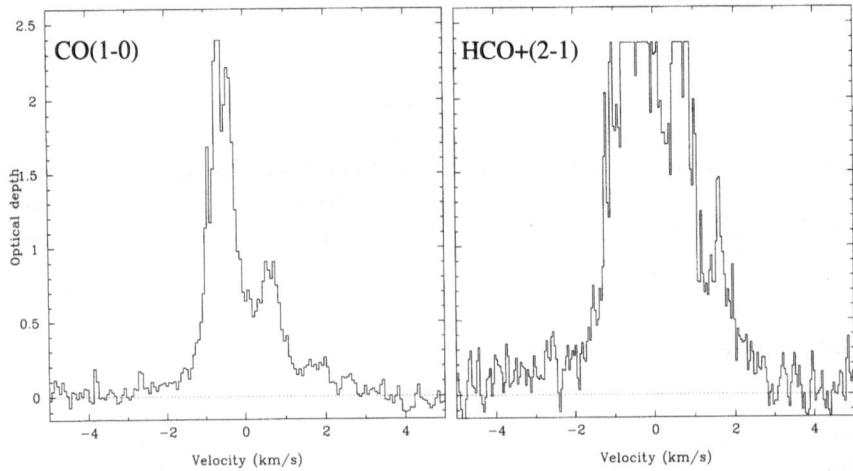

Fig. 4.1. Plots of the observed opacity for the CO(1-0) and HCO$^+$(2-1) transitions seen at $z = 0.25$ towards PKS 1413+135. The cut-off at opacities > 2 are due to saturation of the signals. The opacity of the HCO$^+$(2-1) line is larger than that of the CO(1-0) line despite of an abundance which is $10^{-3} - 10^{-4}$ that of CO.

in column densities is reflected in the large spread in optical extinction, A_V, as well as the atomic to molecular ratio. Systems with high extinction have 10-100 times higher molecular gas fraction than those of low extinction.

a)   b)

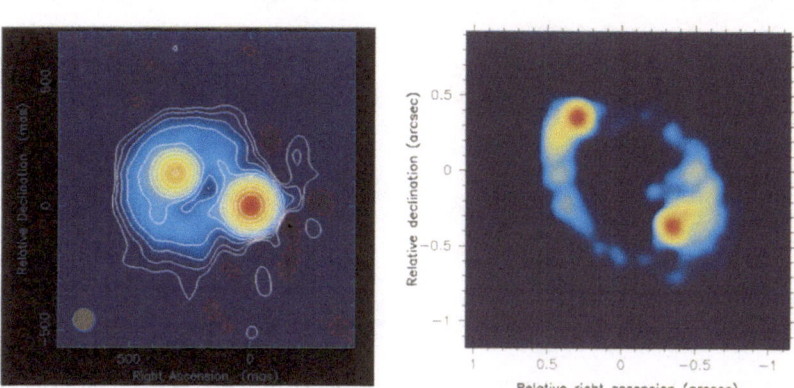

Fig. 4.2. a) A 15 GHz radio image of the gravitational lens B 0218+357 obtained with the VLA (courtesy A. Patnaik). **b)** A 15 GHz radio image of the gravitational lens PKS 1830-211 (see Color Plate).

Absorption in the Host Galaxy. Two of the four known molecular absorption line systems are situated within the host galaxy to the 'background' continuum source: PKS 1413+135 [149] and B3 1504+377 [152]. The latter exhibits two absorption line systems with similar redshifts, z=0.67150 and 0.67335. The separation in restframe velocity is 330 $km\,s^{-1}$. This is the type of signature one would expect from absorption occurring in a galaxy acting as a gravitational lens, where the line of sight to the images penetrate the lensing galaxy on opposite sides of the galactic center. However, in this case, as well as for PKS 1413+135, high angular resolution VLBI images show no image multiplicity, despite impact parameters less than 0.''1 (e.g. [113,162]). The continuum source must therefore be situated within or very near the obscuring galaxy.

Absorption in Gravitational Lenses. The two absorption line systems with the highest column densities occur in galaxies which are truly intervening and each acts as a gravitational lens on the background source: B 0218+357 and PKS 1830-211. In these two systems several isotopic species are detected as well as the main isotopic molecules, showing that the main lines are saturated and optically thick [33,34,151,153]. Nevertheless, the absorption lines do not reach the zero level. This can be explained by the continuum source being only partially covered by obscuring molecular gas, but that the obscured regions are covered by optically thick gas. The lensed images of B 0218+357 and PKS 1830-211 consist of two main components. By comparing the depths of the saturated lines with fluxes of the individual lensed components, as derived from long radio wavelength interferometer observations, the obscuration is found to cover only one of two main lensed components [150,151]. This has subsequently been verified through mm-wave interferometer data [98,153,140].

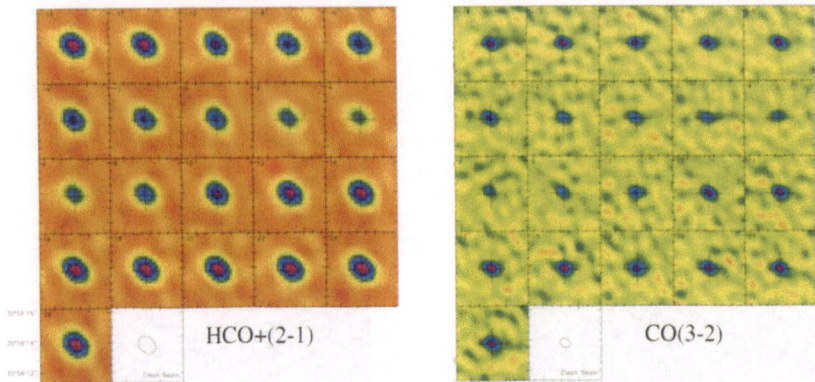

Fig. 4.3. Channel maps of HCO$^+$(2-1) and CO(3-2) absorption towards the B 0218+357 obtained with the IRAM Plateau de Bure interferometer. The angular resolution does not resolve the two lensed images of the background QSO. The continuum weakens in the channel maps which corresponds to the absorption line, but never disappears completely. Since the absorption lines are strongly saturated, this shows that only part of the continuum is obscured by optically thick molecular gas. From Combes & Wiklind (unpublished) (see Color Plate).

B 0218+357. This is a flat-spectrum radio source lensed by an intervening galaxy. The lens nature was first identified by Patnaik et al. [108]. The lens system consists of two components (A and B), separated by 335 milliarcseconds (Fig. 4.2a). There is also a faint steep-spectrum radio ring, approximately centered on the B component. Absorption of neutral hydrogen has been detected at $z_d = 0.685$ [26], showing that the lensing galaxy is gas rich. The redshift of the background radio source is tentatively determined from absorption lines of Mg IIλ2798 and Hγ, giving $z_s \approx 0.94$ [24]. Molecular absorption lines were detected in this system [150] further strengthening the suspicion that the lens is gas-rich and likely to be a spiral galaxy. The molecular absorption lines do not reach zero level. Nevertheless, absorption of isotopic species show that the main isotopic transitions must be heavily saturated. In fact, both the ^{13}CO and C^{18}O transitions were found to be saturated as well, while the C^{17}O transition remained undetected [33] [35]. This gives a lower limit to the CO column density which transforms to $N_{H_2} \approx 4 \times 10^{23}$ cm^{-2} and an $A_V \approx 850$ mag.

That the molecular gas seen towards B 0218+357 covers only one of the two lensed images of the background source can be seen in Fig. 4.3, where the continuum decreases at velocities corresponding to the absorption line but never completely disappears. Subsequent millimeter interferometry observations have shown that the absorption occurs in front of the A-component, which is then expected to be completely invisible at optical wavelengths. Nevertheless, images obtained with the HST WFPC2 in broad V- and I-band, show both components (Fig. 4.4). While the intensity ratio A/B of the two lensed images is 3.6 at radio wavelengths [109], A/B$\approx$0.12 at optical wavelengths. The V$-$I values show no

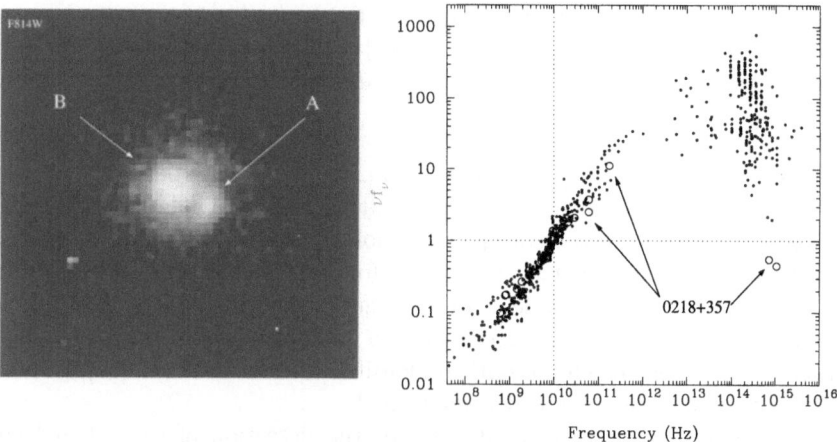

Fig. 4.4. The gravitational lens B 0218+357. **Left :** Optical image obtained with the HST in the I band (FW814), showing the A and B components (archival data). In contrast to the radio image (Fig. 4.2a), the A component is weaker than the B component. This is caused by obscuration of molecular gas, which gives rise to the observed molecular absorption lines at $z_d = 0.688$. **Right :** Normalized SED for flat-spectrum radio QSOs. The fluxes for the A component in B 0218+357 as observed with the HST are marked by circles. Their positions suggest that only 3% of the flux supposed to come from this component reaches the observer. (From [155]).

significant difference in reddening for the A- and B-component. Hence, there is no indication of excess extinction in front of the A-component despite the large A_V inferred from the molecular absorption. Since it is unlikely that the A/B intensity ratio is very much different at optical and radio wavelengths (differential magnification could introduce a small difference if the radio and optical emission comes from separate regions) the A component appears sub-luminous in the optical. The other possibility is that the B component is over-luminous at optical wavelengths by a factor 30 (or 1.4 magnitudes), possibly caused by microlensing. This latter explanation is, however, quite unlikely in view of the presence of large amounts of obscuring molecular gas in front of the A component. By compiling a sample of flat-spectrum radio sources from the literature, with properties similar to that of B 0218+357 (except the gravitational lensing aspect), correcting for different redshift and normalizing the observed luminosities at $\nu = 10$ GHz, it is possible to show that the optical luminosity of B 0218+357 is abnormally weak [155] (Fig. 4.4). In this comparison the observed magnitude of the A component was used, multiplied by a factor 1.3 in order to compensate for the B component using the magnification ratio of 3.6. This clearly showed the A component to be sub-luminous, rather than the B component being over-luminous. The interpretation of this is that the A component is obscured by molecular gas, with an extinction that is very large. Some light 'leaks' out but through a line of sight which contains very little obscuring gas, hence not showing much reddening in the V−I colors. Assuming that all the obscuration occurs in the A

component, only ∼3% of the photons expected from the A component reaches
the observer. Since the extent of the optical emission region is very small, this
suggests the presence of very small scale structure with a large density contrast
in the molecular ISM of the lensing galaxy.

PKS 1830-211. This is a radio source consisting of a flat-spectrum radio core
and a steep-spectrum jet. It is gravitationally lensed by a galaxy at $z_d = 0.886$
[151] into two images of the core-jet morphology (Fig. 4.2b). The two cores are
separated by $0''.97$ and the images of the jet form an elliptical ring. PKS 1830-211
is situated close to the Galactic center and suffers considerable local extinction.
Its lens nature was first suspected through radio interferometry [120], but as
neither redshift was known nor optical identification achieved (cf. [39]) its status
as a gravitational lens remained unconfirmed.

The lensing galaxy was found through the detection of several molecular
absorption lines at $z_d = 0.886$ [151]. At millimeter wavelengths the flux from
the steep-spectrum jets is completely negligible and it is only the cores that
contribute to the continuum. It was soon found that the molecular absorption
was seen only towards one of the cores, the SW image. However, weak molecular
absorption was subsequently found also towards the NE image. This fortunate
situation gives two sight lines through the lens and gives velocity information
which can be used in the lens modeling (see Sect. 4.6). A second absorption line
system has been found towards PKS 1830-211, seen as 21cm HI absorption at
$z = 0.19$ [93], making this a possible compound lens system. This intervening
system complicates the lens models of this system. A potential candidate for the
$z = 0.19$ absorption has been found in HST NICMOS images [87]. It is situated
∼4″ SW of PKS 1830-211 and is designated as G2. The molecular absorption
lines towards PKS 1830-211 and their use for deriving the differential time delay
between the two cores will be described in more detail in Sect 4.5.3 and Sect. 4.6.

4.4   Dust Continuum Emission

The spectral shape of the far-infrared background suggests that approximately
half of the energy ever emitted by stars and AGNs has been absorbed by dust
grains and then re-radiated at longer wavelengths [118,52,83,57]. The dust is
heated to temperatures of 20-50 K and radiates as a modified black-body at far-
infrared wavelengths. At the Rayleigh-Jeans part of the dust SED the observed
continuum flux increases with redshift. This is known as a 'negative K-correction'
and is effective until the peak of the dust SED is shifted beyond the observed
wavelength range, which occurs at $z > 10$. Dust continuum emission from high
redshift objects is therefore observable at millimeter and submillimeter wavelengths and is an important source of information about galaxy formation and
evolution in general and for gravitational lenses in particular.

4.4.1 Dust Emission

Dust grains come in two basic varieties, carbon based and silicon based. Their size distribution ranges from tens of microns down to tens of Ångströms. The latter are known as PAH's (Polycyclic Aromatic Hydrocarbonates). Except for the smallest grains, the dust is in approximate thermodynamical equilibrium with the ambient interstellar radiation field. The dust grains absorb the photon energy mainly in the UV and re-radiate this energy at infrared and far-infrared (FIR) wavelengths. The equivalent temperature of the dust grains amount to 15-100 K and they emit as an approximate blackbody.

The spectral energy distribution (SED) of dust emission is usually represented by a modified blackbody curve, $F_\nu \propto \nu^\beta B_\nu(T_d)$ (cf. [142,157]), where B_ν is the blackbody emission, T_d the dust temperature and ν^β is the frequency dependence of the grain emissivity, which is in the range $\beta = 1 - 2$. Such representations have successfully been used for cold dust components where a large part of the SED is optically thin. When $\tau \approx 1$ or larger, the observed dust emission needs to be described by the expression:

$$F_\nu = \Omega_s B_\nu(T_d) \left(1 - e^{-\tau_\nu}\right) , \qquad (4.2)$$

where Ω_s is the solid angle of the source emissivity distribution, τ_ν is the opacity of the dust. Setting $\tau_\nu = (\nu/\nu_0)^\beta$ gives $F_\nu \propto \nu^\beta B_\nu(T_d)$ for $\tau_\nu \ll 1$ and $F_\nu \propto B_\nu(T_d)$ for $\tau_\nu \gg 1$. The critical frequency ν_0 is the frequency where $\tau_\nu = 1$.

The Infrared Luminosity. The total infrared luminosity is derived by integrating (4.2) over all frequencies. Here the flux density F_ν corresponds to the energy emitted by dust only. The infrared luminosity for an object at a redshift z is given by

$$L_{IR} = 4\pi (1 + z)^3 D_A^2 \int_0^\infty F_\nu d\nu , \qquad (4.3)$$

where D_A is the angular size distance[3]. The solid angle Ω appearing in (4.2) is a parameter derived in the fitting procedure. In the event of a single dust component, Ω can be estimated from the measured flux F_{ν_r} at a given restframe frequency ν_r

---

[3] Expressing (4.3) in a form directly accessible for integration, we get

$$\frac{L_{IR}}{L_\odot} = 8.53 \times 10^{10} (1 + z)^3 \left[\frac{D_A}{Mpc}\right]^2 T_d^4 \Omega \int_0^\infty \frac{x^3 \left(1 - e^{-(ax)^\beta}\right)}{e^x - 1} dx .$$

The integral can be integrated numerically with appropriate values of the parameter $a = kT_d/h\nu_0$.

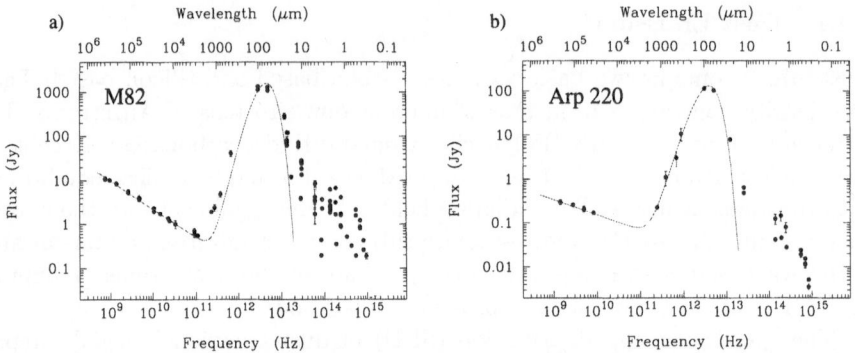

Fig. 4.5. The spectral energy distribution of two starburst galaxies. **a)** M82 and **b)** Arp220. Despite a difference in far-infrared luminosity of almost 2 orders of magnitude, their spectral energy distribution are nearly identical. Notice also the presence of cold dust in M82, visible as an excess flux at millimeter and submillimeter wavelengths. The SEDs have been fitted by a modified blackbody curve, which becomes optically thick at 50μm and which has $\beta = 2.0$ for M82 and $\beta = 1.3$ for Arp220 and using a single temperature component of $T_d = 45$ K for both galaxies.

$$\Omega = \frac{F_{\nu_r}}{B_{\nu_r}(T_d)\left(1 - e^{-(\nu_r/\nu_0)^\beta}\right)}$$

$$\approx 6.782 \times 10^{-4} \left[\frac{\nu_r}{\text{GHz}}\right]^{-3} \left[\frac{F_{\nu_r}}{\text{Jy}}\right] \left[\frac{e^{h\nu_r/kT_d} - 1}{1 - e^{-(\nu_r/\nu_0)^\beta}}\right] . \tag{4.4}$$

Using some typical values ($T_d = 30$ K, $\beta = 1.5$, $\nu_0 = 6$ THz (50μm), $\nu_r = \nu_{obs}(1+z) = 1400$ GHz ($\nu_{obs} = 350$ GHz at $z = 3$) and, finally, an observed flux of 1 mJy) we get $\Omega \approx 2 \times 10^{-14}$. For a spherical source with a radius $r \approx D_A\sqrt{\Omega/\pi}$, this corresponds to a dust continuum emission region with an extent of only ~ 160 pc.

Although this is a very rough estimate of the size of the emitting region, it shows, since typical observed values were used, that FIR dust emission from distant objects tend to come from very small regions. This will be of importance when considering the effects of gravitational lensing.

The Dust Mass. An estimate of the dust mass from the infrared flux requires either optically thin emission combined with a knowledge of the grain properties, or optically thick emission and a knowledge of the geometry of the emission region (cf. [64]).

The grain properties are characterized through the macroscopic mass absorption coefficient, κ_ν. Several attempts to estimate the absolute value of κ_ν as well as its frequency dependence have given different values (cf. [66]). Combining the same frequency dependence as in [66] with a value given by [64], the mass absorption coefficient can be described as

$$\kappa_{\nu_r} \approx 0.15 \left(\frac{\nu_r}{375\text{GHz}}\right)^{1.5} \text{m}^2 \text{ kg}^{-1} , \tag{4.5}$$

where ν_r corresponds to the restframe frequency. This expression corresponds to a grain composition similar to that found in the Milky Way. At frequencies where the emission is optically thin, the dust mass can now be determined from[4],

$$M_d = \frac{F_{\nu_{obs}}}{\kappa_{\nu_r} B_{\nu_r}(T_d)} D_A^2 (1+z)^3 . \tag{4.6}$$

4.4.2 Detectability of Dust Emission

A typical far-infrared spectral energy distribution (SED) of a starburst galaxy (M82) is shown in Fig. 4.5a. The SED of a more powerful starburst (Arp220) is shown in Fig. 4.5b. Perhaps the most striking aspect of these SEDs is their similarity, despite that they represent galaxies with widely different bolometric luminosities. In both cases most of the bolometric luminosities comes out in the far-infrared: M82 has a far-infrared luminosity of $3 \times 10^{10} \, L_\odot$, while Arp220 is a so called Ultra-Luminous Infrared Galaxy (ULIRG) with a far-infrared luminosity of $1 \times 10^{12} \, L_\odot$. The SEDs shown in Fig. 4.5 have been fitted by a modified blackbody curve, which becomes optically thick at $50\mu m$ and which has $\beta = 2.0$ for M82 and $\beta = 1.3$ for Arp220 (cf. (4.2)). The modified blackbody curves has been fitted using a single temperature component of $T_d = 45$ K for both galaxies. Notice, however, the presence of a colder dust component in the SED of M82, which is visible as an excess flux at millimeter and submillimeter wavelengths [143].

The observed dust continuum emission originates from dust grains in different environments and which are heated by different sources. Nevertheless, a remarkably large number of dust SEDs, like the ones shown in Fig. 4.5, can be well fitted by only one, or in some cases two dust components (cf. the cold dust component in M82).

For a single dust temperature component, the flux ratio between the submillimeter $(850\mu m)$ and the far-infrared $(100\mu m)$ is strongly dependent on the dust temperature. For $T_d = 45$ K (as in the case of M82 and Arp220), $f_{850\mu m}/f_{100\mu m} \approx 3 \times 10^{-3}$, while for $T_d = 20$ K, $f_{850\mu m}/f_{100\mu m} \approx 0.06$, or 20 times larger. Nevertheless, as long as the dust temperature is not extremely low, it is much harder to observe the long wavelength tail of the dust SED than the peak at $\sim 100\mu m$ (except that in the latter case one needs to observe from a satellite due to our absorbing atmosphere).

At millimeter and submillimeter wavelengths the SED can, to a first approximation, be characterized by $f_\nu \propto \nu^\gamma$, where $\gamma = 3-4$. Hence, the observed flux increases as an object is shifted to higher redshift. This effect is large enough to completely counteract the effect of distance dimming. An example of this is

---

[4] (4.6) can also be expressed as:

$$M_d \approx 4.08 \times 10^4 \times \left[\frac{F_{\nu_{obs}}}{Jy} \right] \left[\frac{D_A}{Mpc} \right]^2 \left(\frac{\nu_r}{375 GHz} \right)^{-9/2} \left(e^{h\nu_r/kT_d} - 1 \right) (1+z)^3 \, M_\odot .$$

shown in Fig. 4.6, where the observed flux at $850\mu m$ has been calculated for a FIR luminous, 5×10^{12} L$_\odot$, galaxy, for two different dust temperatures, $T_d = 30$ K and $T_d = 60$ K, and for two different cosmologies. The largest uncertainty in the predicted flux as a function of redshift comes from the assumed dust temperature, rather than the assumed cosmology. However, regardless of dust temperature and cosmology, the effect of the 'negative K-correction' of the dust SED is to make the observed flux more or less constant between redshifts of $z = 1$ and all the way to $z \approx 10$, where the Wiener part of the modified blackbody curve is shifted into the submillimeter window and the flux drops dramatically.

This constant flux over almost a decade of redshift range makes the millimeter and submillimeter window extremely valuable for studies of the formation and evolution of the galaxy population at high redshift in general and for gravitational lensing in particular. For a constant co-moving volume density, the submm is strongly biased towards detection of the highest redshift objects. The prerequisite is, of course, that galaxies containing dust exist at these large distances and that the low flux levels expected can be reached by our instruments. Both of these criteria are actually fulfilled; powerful new bolometer arrays working at millimeter (MAMBO, and recently SIMBA) and in the submillimeter (SCUBA) have shown that low flux levels can be observed and that objects containing large amounts of dust do exist at early epochs (cf. [67,134,44]).

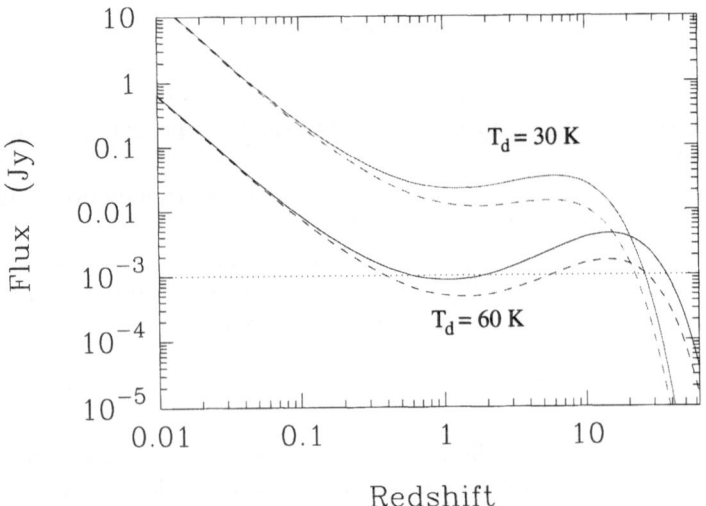

Fig. 4.6. The observed flux density at $850\mu m$ of a source with $L_{\rm FIR} = 5 \times 10^{12}$ L$_\odot$ as a function of redshift. The top set of curves correspond to a dust temperature of 30 K and the bottom curves to a dust temperature of 60 K. The full drawn lines correspond to a flat matter dominated universe ($\Omega_m = 1.0, \Omega_\Lambda = 0$), and the dashed curve to a flat Λ dominated universe ($\Omega_m = 0.3, \Omega_\Lambda = 0.7$).

4.4.3   Submillimeter Source Counts

One of the first studies using long wavelength radio continuum emission was to simply count the cumulative number of detected sources as a function of flux level. These observations mainly probed high luminosity radio galaxies and showed a significant departure from an Euclidean non-evolving population. This was the first evidence of cosmic evolution [122,74].

The negative K-correction in the mm-to-far-infrared wavelength regime for dust emission has enabled present day submm/mm telescopes, equipped with state-of-the-art bolometer arrays, to get a first estimate of the source counts of FIR luminous sources at high redshift. There are two bolometer arrays which have produced interesting results so far; SCUBA on the JCMT in Hawaii and MAMBO on the IRAM 30m telescope in Spain. Additional arrays are under commissioning and will likely contribute to this area shortly: SIMBA on the 15m SEST on La Silla, and BOLOCAM on the 10m CSO on Mauna Kea.

The SCUBA bolometer array at the JCMT was put to an ingenious use when it looked at blank areas of the sky chosen to be towards rich galaxy clusters at intermediate redshift [132,134,6,18,19]. The gravitational magnification by the cluster enabled very low flux levels to be reached and several detections were reported. This method has been used by others as well and an example of an image of the rich cluster Abell 2125 at $1250\mu m$ is shown in Fig. 4.7 [30]. More than a dozen sources are detected above the noise but none is associated with the cluster itself. Instead they are all background sources gravitationally magnified by the cluster potential.

The cumulative source count of a population of galaxies is simply the surface density of galaxies brighter than a given flux density limit. In a blank field observation it is in principle derived by dividing the number of sources with the surveyed area. The effects of clustering have to be considered if the observed area is small. In practice there are several statistical properties that have to be considered. Usually the threshold for source detection is not uniform across the mapped area. Since the sources are generally found close to the detector limit, those which have fluxes boosted by spurious noise have a higher likelihood to be detected than those which experience a negative noise addition, which are likely to be lost from the statistics. This latter effect leads to an overestimate of the true source flux. The possibility of an additional bias through differential magnification will be discussed in Sect. 4.4.5.

The case of submm/mm detected galaxies behind foreground galaxy clusters is yet more complicated (cf. [18]). The gravitational lens distorts the background area and magnifies the source fluxes. The magnitude of these effects may vary across the observed field. A detailed mass model of the lens is needed in order to transform the observed number counts into real ones, as well as knowledge about the redshift distribution of the sources. Smail and collaborators ([132–134]) initially observed 7 clusters, constructed or used existing mass models of the cluster potentials, and managed to obtain source counts at sub-mJy levels (cf. [18]). Although the lensing effect of clusters allows observations of weaker fluxes, it introduces an extra uncertainty in the number counts. This is, however,

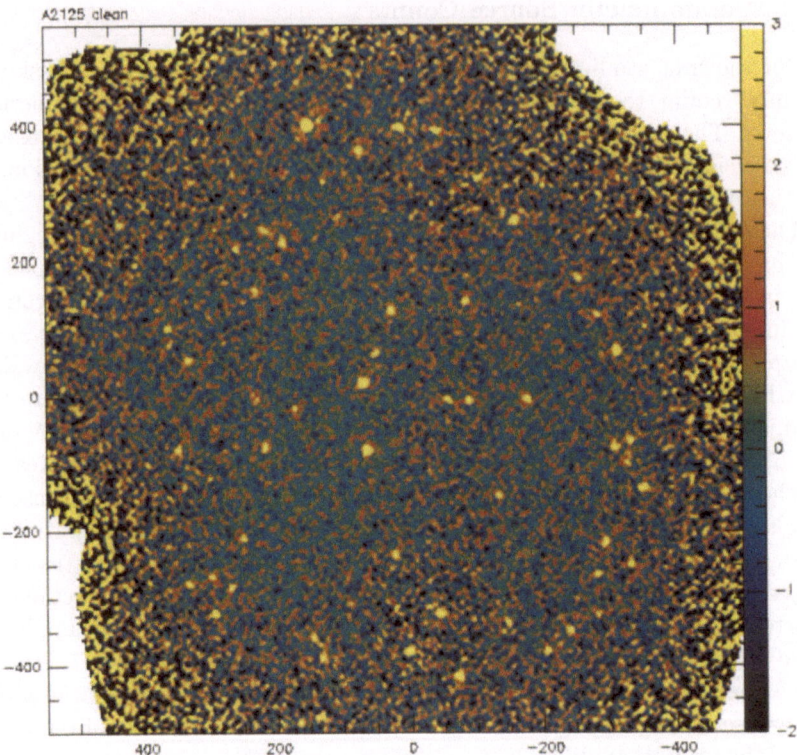

Fig. 4.7. An image of the cluster Abell 2125 obtained with the MAMBO bolometer array at the IRAM 30m telescope (from [30]). The angular size is in arcseconds. The noise rms is at 0.5 mJy/beam (see Color Plate).

not dominating the overall error budget [18]. There is another beneficial effect with the lensing in that the extension of the background area alleviates the problem of source confusion. The angular resolution of existing bolometer arrays is approximately 15″ and source confusion is believed to be a problem at flux levels below 0.5 mJy.

Other blank field surveys using SCUBA have pushed as deep as the cluster surveys, but without the extra magnification they probe somewhat higher flux levels. Examples of such deep blank field surveys include the Hubble Deep Field North [67], the fields used for the Canada-France Redshift Survey [44,45], the Lockman hole and the Hawaii deep field region SSA13 [5].

All these submm deep fields, including the cluster fields, are only a few square arcminutes. Using on-the-fly mapping techniques a few groups have recently started mapping larger areas but to a shallower depth (cf. [21] [131]).

Carilli et al. [30] combined the number counts from all the blank-field observations. The result is a cumulative source count stretching from ∼15 mJy to 0.25 mJy (Fig. 4.8). The source counts obtained using the lensing technique, after correcting for the lensing effects, are compatible with those obtained through

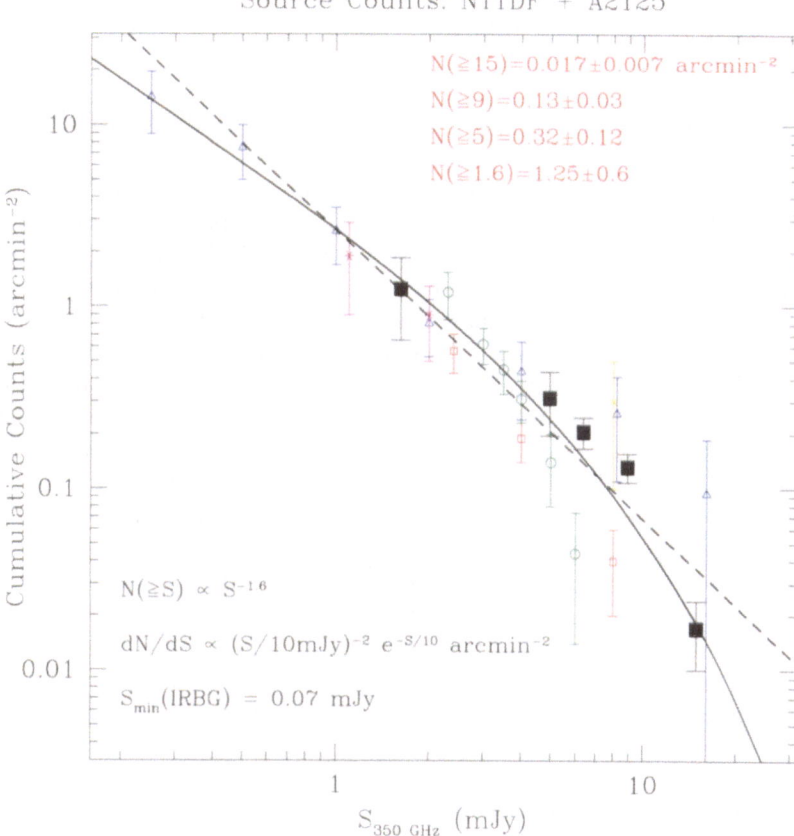

Fig. 4.8. Source counts from several surveys using SCUBA at 850μm and MAMBO at 1250μm (from [30]). The dashed curve is a powerlaw of index -1.8 while the solid curve is an integrated Schechter luminosity function with a powerlaw index -2 and an exponential cut-off at 10mJy. All fluxes refer to 850μ. The 1250μm data points have been multiplied by a factor 2.25 in order to transform them into expected fluxes at 850μm (see Color Plate).

pure blank-fields. The turnover at a flux level of $\sim$10 mJy is probably real and represents a maximum luminosity of $\sim 10^{13}$ L$_\odot$ for an object at $z \approx 3$. The exact shape of the number counts is still uncertain at both the low and high flux ends. Results from the MAMBO bolometer array, which operates at 1250μm, have been multiplied by a factor 2.25 in order to transform it into the expected flux at 850μm. This assumes that the objects have an SED of the same type as starburst galaxies (cf. Fig. 4.5).

In order to transform the cumulative source count into a volume density it is necessary to know the redshift distribution of the sources. It is, however, possible to circumvent this by fitting a model of galaxy evolution to the observed source counts. This has been explored extensively by Blain et al. [19], (see also [31] [141]), and will not be discussed further here.

4.4.4   Submm Source Identification and Redshift Distribution

The sources detected in submm/mm surveys can in a majority of cases be identified with sub-mJy radio sources (cf. [135]). This population of weak radio continuum sources is believed to be powered by star formation rather than AGN activity [161,62]. Attempts to identify the submm/mm sources with optical and/or infrared counterparts have failed in all but a small number of cases (cf. [42,72,55]). The submm/mm detected population is not related to nearby nor intermediate redshift sources, but are believed to be at $z > 1$, but the lack of clear optical/IR identifications has made it difficult to assess its true redshift distribution. An alternative technique for determining the redshift has been introduced by Carilli & Yun [29], which relates the radio continuum flux at 1.4 GHz with the measured flux at $850\mu m$ (see also [7]). As the radio flux declines with increasing redshift, the submm flux increases (cf. Fig. 4.5). Although the method is model dependent (mainly depending on the dust temperature T_{dust}, the radio spectral index as well as the frequency dependence of the dust emissivity coefficient, cf. Sect. 4.4.1), it gives a rough estimate of the redshift. Using this method it has been possible to show that the majority of the submm/mm detected sources lie at a redshift $1 \leq z \leq 4$ (cf. [135,30]).

4.4.5   Differential Magnification

One well-known property of gravitational lensing is that it is achromatic, meaning that the deflection of photons by a gravitational potential is independent of wavelength. The achromaticity is applicable to observed gravitational lenses as long as the source size is small compared to the caustic structure of the lens, such as when the Broad Line Region (BLR) of a QSO is lensed by a galaxy sized lens. Chromatic effects can, however, become important if the source is substantially extended (relative to the caustic structure) and the spectral energy density of the source is position dependent.

The submm/mm detected dusty sources discussed in Sect. 4.4 are characterized by extended emission, several orders of magnitude larger than the compact sources generally studied in gravitational lensing. This applies to dust emission regardless whether the dust is heated by star formation or by a central AGN. Measured on galactic scales, however, the dust is relatively centrally concentrated, with typical scales ranging from 10^2 pc to a few kpc (cf. Sect. 4.4.1). A dust distribution heated by a central AGN will have a radial dust temperature distribution, even when radiation transfer effects and a disk- or torus-like geometry are considered. This is observed in nearby Seyfert galaxies [115]. A radial temperature profile is also found in the case of a pure starburst [130], but spatially more extended than in the AGN case. Gravitational lensing of an extended

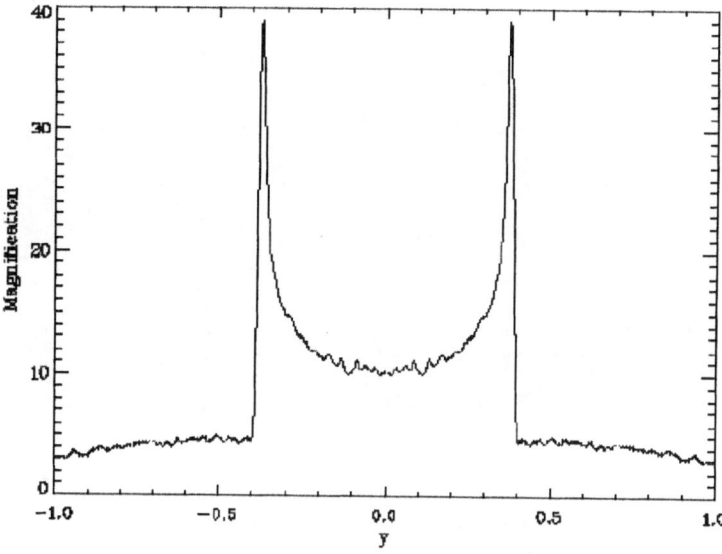

Fig. 4.9. Magnification in a cut through the caustic structure of an elliptical lens configuration. The two peaks corresponds to the radial caustic. This particular cut does not pass through the tangential caustic, which would have produced a yet stronger magnification peak close to the center. The length scale is in arcseconds. Notice the very strong gradient in the magnification when going from the one-image region to the the extended three-image 'plateau'. The magnification changes by a factor ~ 10 over angular scales of $\sim 0''01$, corresponding to scales of ~ 50 pc at $z_s \approx 2$. From K. Pontoppidan's Master Thesis, Copenhagen University [116].

dust distribution with a non-homogeneous temperature, and thus emissivity distribution, means that the assumption of achromaticity is no longer valid and the source may be differentially magnified.

If the characteristic length scale in the source plane is η_0 the characteristic length scale in the lens plane is $\xi_0 = (D_d/D_s)\,\eta_0$. Taking a dust distribution of 1 kpc (η_0), a source redshift $z_s = 3$ and a lens redshift $z_d = 1$, the characteristic length scale in the lens plane becomes approximately $0''2$. This is close to the typical image separation for strong lensing. Since the submm/mm detected galaxies are believed to show a significant change in the dust temperature over this scale, it is quite likely that they will exhibit chromatic effects.

An analytical model of the effect of differential magnification of dusty sources was presented in [16], where it was shown that the effect can be strong and that it was most likely to produce an increase in the mid-infrared flux relative to the long wavelength flux. This would make the sources appear warmer than what their intrinsic SED would imply.

A more detailed analysis of the effect of differential magnification and its probability for occurance was done by Pontoppidan [116]. Using elliptical potentials and a realistic parameterization of the dust and its spectral energy distri-

size of dust region 100 pc   size of dust region 500 pc

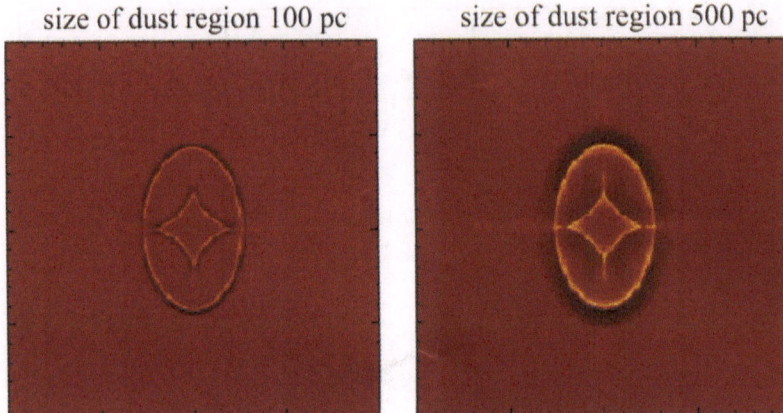

Fig. 4.10. The caustic structure of an elliptical potential showing areas where the dust spectral energy distribution will be influenced by differential magnification. The dust emission region is modeled as a circular disk with a radial temperature profile. The central heating can be either an AGN or a dense starburst. If the center of the dust emission region is placed in the black area, the observed SED will appear cooler than the intrinsic one, while the opposite effect occurs if the center is located in the black areas. The effect of a small (left) and large (right) dust region is illustrated. From K. Pontoppidan's Master Thesis, Copenhagen University [116] (see Color Plate).

bution it was showed that both positive and negative distortions of the SED can occur. Here positive means an increase in the mid-IR part and negative means an increase in the far-IR/submm part. Fig. 4.9 shows a plot of the magnification in a cut through one of these models (elliptical potential) which does not hit the inner tangential caustic. By placing the center of a dust emission region, with a radial temperature profile, well outside the radial caustic (i.e. $> |0\overset{''}{.}5|$ from the center), parts of the outer region of the dust distribution will fall on the high magnification plateau inside the radial caustic and be multiply imaged while the center is singly imaged and only moderately magnified. This situation would cause an enhancement of the long wavelength part of the SED relative to the mid-IR part. The radius of a typical dust distribution is $\sim 0\overset{''}{.}2$ at $z \approx 3$. If the center of the source is placed closer to the radial caustic, both the center and the extended dust distribution will be magnified, but the warmer central dust will experience a larger average magnification and hence result in a flattening of the SED at mid-IR wavelengths. Again, it might be that the cool dust is multiply imaged while the center (possibly containing an AGN) is singly imaged.

Pontoppidan [116] found that the cross section for an enhancement of the long wavelength part of the SED is larger than for an enhancement of the mid-IR. However, the latter situation results in a stronger magnification and effects the observed SED to a higher degree. The latter case also represents a situation where the system is more likely to be recognized as a gravitational lens.

Effect on Number Counts of submm/mm Detected Galaxies. A distortion map of the effect of differential magnification is shown in Fig. 4.10 (from [116]). The caustic structure of an elliptical potential representing a central mass surface density of 3×10^9 $M_\odot$ kpc^{-2}, where the distortion of the dust SED due to differential magnification has been color coded. Black represents a negative distortion (cooler SED) and white represents a positive distortion (warmer SED). The dust distribution of the source is assumed to have a radius of 100 pc (left image) and 500 pc (right image). Quite naturally, the larger the region over which dust is distributed, the larger is the region where negative distortion can occur. By placing the center of the source in the black/white regions of the distortion map, the observed SED will appear cooler/warmer.

The implications for the submm/mm detected objects at high redshift is that differential magnification could induce a bias in the number counts. Especially since in most of the surveys done so far the sources are found close to the detection limit. The effect could induce an overestimate of the number of sources but it could also influence the slope of the cumulative number counts. The latter is more likely but better statistics from surveys reaching low noise levels are needed, as well as a better understanding of the total cross section for positive/negative distortions of dusty submm/mm sources.

An interesting consequence of the differential magnification of these sources is that some, perhaps several, of the detected submm/mm objects may be multiply imaged systems when viewed at high angular resolution in submm/mm wavelengths. The radio identifications that have been done typically reach a resolution of $1''$ which is not sufficient to see multiple images on the expected $0''.1$-$0''.2$ scale. High angular resolution deep imaging with future instruments such as ALMA will resolve this issue.

4.5 Case Studies

In order to describe in more detail the characteristics of millimeter observations and interpretations of gravitationally lensed sources, as well as to illustrate their use, three cases are presented below. First is the luminous Broad-Absorption-Line (BAL) quasar APM 08279+5255 at $z = 3.9$. The gravitational lens hypothesis for this source was put forward based only on its apparent luminosity. The second case is a detailed study of the quadruply lensed Cloverleaf quasar, where the gravitational lensing of molecular gas has enabled a more detailed and constrained lens model. The last example is PKS 1830-211, where the lensing galaxy was actually first detected through millimetric molecular absorption lines at $z_d = 0.886$. The molecular absorption lines in this system has been used to constrain the lens model by giving the velocity dispersion and are used to derive the differential time delay between the two main lensed components.

4.5.1 APM 08279+5255: A Case of Differential Magnification?

This object was discovered serendipitously during a search for Galactic carbon stars [70]. It was found to be a BAL QSO at a redshift $z = 3.911$ (see [43] for the redshift determination). With an astounding R-band magnitude of 15.2 and detection in three of the four IRAS bands, its bolometric luminosity turns out to be 5×10^{15} $L_\odot$. This in itself led to the suspicion that it is a gravitationally lensed object Subsequent observations, both from the ground and from space [85,47,69], led to the detection of three components, with a maximum separation of $0\overset{''}{.}35 \pm 0\overset{''}{.}02$, and with a flux ratio of the two brightest components of 1.21 ± 0.25 (cf. Fig. 4.13). The optical spectra of the two main components are similar to each other [85]. No lensing galaxy has been identified, although the weak third image could potentially be the lens (see below). Nevertheless, based on the small separation of the main components, their similar spectra and the enormous luminosity inferred for the system, the lensing nature of this system is not questioned. Even in the case of strong gravitational magnification, APM 08279+5255 is an intrinsically very luminous system, with $L_{bol} \geq 10^{13}$ $L_\odot$.

The high apparent brightness of APM 08279+5255 has allowed a very good S/N optical spectra of the intervening absorption line systems to be obtained with the HIRES spectrograph on Keck [48]. Several potential lens candidates are found as MgII absorption line systems, with the most conspicuous one at $z = 1.181$. Placing the third image at this redshift, however, requires the lens to be unusually compact and luminous. It would need to be almost 5 magnitudes brighter than an $L^\star$ galaxy with the relevant velocity dispersion of ~ 150 km s^{-1} [69,47]. The possibility that the lens harbors an AGN can be dismissed since no emission lines from $z < 3.9$ are detected in the spectrum. Also, the continuum of an intervening QSO should have been detected in the saturated parts of the absorption lines seen towards the background source. No such emission is detected (cf. [48]). APM 08279+5255 could thus represent a 'text book' example of a gravitational lens with an odd number of components.

Apart from being luminous in the optical and UV, APM 08279+5255 also contains large amounts of dust and metal rich molecular gas (Fig. 4.11). The SED of APM 08279+5255 is actually dominated by a strong dust continuum emission (Fig. 4.12), detected over a wide wavelength band: from the restframe submm to mid-infrared bands. This puts APM 08279+5255 in the class of hyperluminous IR galaxies even when correcting for a strong gravitational magnification.

The overall dust spectral energy distribution is characterized by a steeply rising long wavelength part, with a change of slope around $\lambda_{rest} = 200\mu m$, and a flat mid-IR part. The dust continuum spectra can be fitted by two dust components. One 'cool' characterized by a dust temperature of $T_d = 200$ K, which is optically thin at $\lambda > 200\mu m$ (cf. Fig. 4.12). The second component is hot, with $T_d \approx 910$ K, close to the sublimation temperature of carbon based dust grains. This second component is optically thick. The total dust mass, uncorrected for gravitational magnification, is 2×10^8 $M_\odot$, most of it contained in the cool dust component.

The CO emission lines shown in Fig. 4.11 include the high excitation transition $J = 9 - 8$. The CO $J = 9$ level is $J(J + 1) \times 2.77 = 249$ K above the ground state. Normal type Galactic molecular clouds with typical H_2 densities of $\sim 300 - 10^3$ cm^{-3} are not sufficient to collisionally populate the CO

$J = 9$ level. The mere detection of the CO(9-8) line therefore shows that the gas has to be unusually dense and warm. This immediately suggests that this gas component resides close to the QSO, possibly associated with the hot dust component. If both the CO $J = 4 - 3$ and $J = 9 - 8$ emission are associated with the same gas component, the total molecular gas mass, corrected for magnification, is quite modest: 3×10^9 M$_\odot$ [43]. If the lower transition, on the other hand, emanates from a more extended and cooler region than the $J = 9 - 8$ transition, the total molecular gas mass can be one to two orders of magnitude larger. That this is likely to be the case was shown by the detection of CO $J = 1 - 0$ and $J = 2 - 1$ emission from APM 08279+5255 [106][5]. Using the same conversion factor between H$_2$ column density and velocity integrated CO intensity as is used for the Milky Way and nearby galaxies, the total molecular gas mass in APM 08279+5255, uncorrected for gravitational magnification, is $(0.6 - 3.2) \times 10^{11}$ M$_\odot$ [106]. The amount of gravitational magnification is in this case expected to be low due to the extended nature of the molecular gas, especially the gas seen in the lower transitions. Incidentally, three additional CO emitting sources are detected within $3''$ of the center of APM 08279+5255 [106]. If these are not gravitationally lensed images, which they are not if the currently best lens models are used (cf. [69,47]), these three additional sources are not magnified to any significant degree. The field around APM 08279+5255 should then represent a remarkable over-density of gas rich galaxies at high redshift. These three additional sources are, however, not detected in continuum emission with the Plateau de Bure interferometer nor at optical or NIR wavelengths and their exact nature remains undetermined.

APM 08279+5255 has a SED which is essentially flat from a restframe wavelength of $\sim$30μm to optical wavelengths (cf. Fig.4.12). This is usually interpreted as being the effect of a face-on configuration of a dust-disk surrounding a central AGN. The low inclination of the disk enables the observer to get an un-obscured view of the hot dust close to the AGN as well as the cool dust further away. Comparison with dust models calculated by [58,59] shows that the mid-IR slope is too shallow even for the most extreme face-on models [89], i.e. the dust SED in APM 08279+5255 appears to be too 'warm' even if heated by a powerful AGN. Another possible explanation for the flat mid-IR SED is that APM 08279+5255 experiences differential magnification of the dust emission region. This possibility was explored by [47] by applying sources of various sizes to their lens model. In Fig. 4.12 the SED of APM 08279+5255 is shown together with a starburst model [121]. The starburst model has been arbitrarily fitted to the long wavelength part of the observed SED. At mid-IR wavelengths, the starburst model predicts a flux which is $\sim$50 times lower than the observed fluxes in APM 08279+5255. The data points marked by open circles are the equivalent observed fluxes diminished by a factor of 50 in order to fit the starburst model. Although APM 08279+5255

---

[5] The offset between the CO emission presented in [106] and that of [43] results from the use of slightly different coordinates for APM 08279+5255. The coordinates given in the caption of Fig. 4.11 corresponds to the best optical/IR coordinates determined from both ground and space based imaging.

undeniably contains a powerful AGN, which is likely to contribute a substantial part of the heating of the gas and dust, the influence of star formation can not be ruled out. Can differential magnification account for at least part of the difference between a pure starburst SED and the observed one?

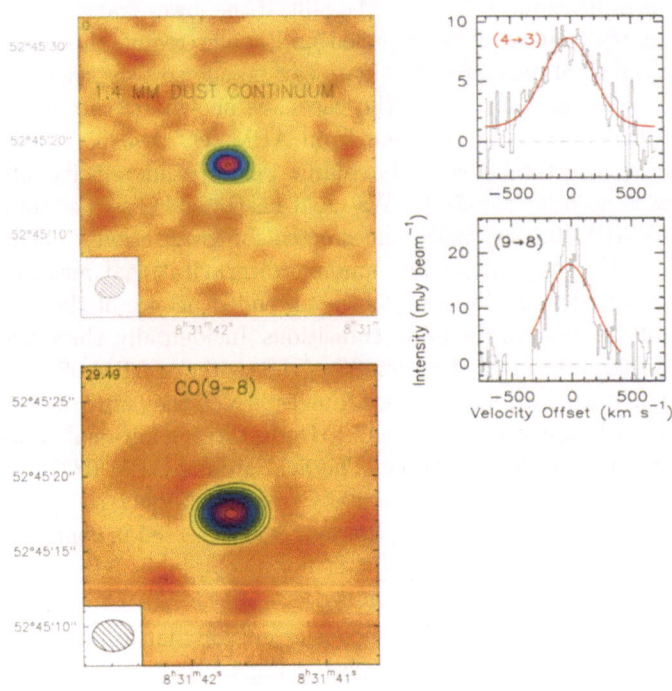

Fig. 4.11. CO spectra and maps of APM 08279+5255 observed with the IRAM Plateau de Bure interferometer (Downes et al. [43]). **Upper left:** 1.4mm dust continuum. **Lower left:** CO(9-8) emission. **Upper and lower right:** CO(4-3) and CO(9-8) emission line profiles. The angular resolution of the maps is $3\rlap{.}''2 \times 2\rlap{.}''3$, far too coarse to resolve the individual components seen at optical/NIR wavelengths. The maps are centered on $08^h31^m41\rlap{.}^s70$, $+52°45'17\rlap{.}''35$ (J2000) (see Color Plate).

Modeling the Lens APM 08279+5255. The lensing configuration of this system has been modeled by Egami et al. [47] and Ibata et al. [69]. Using an isothermal elliptical potential with no external shear, two different types of lens models were applied: a three-image model and a two-image model. The former model is non-singular in order to produce the third image, while the latter assumes that the third image is the lensing galaxy and the potential is singular in order to suppress the formation of the third image. The two-image model produce a modest magnification of $\sim$7 (cf. [47]), while the three-image model produce a magnification of $\sim$90 for both a point source and a more extended source distribution [69,47]. Since the apparent bolometric luminosity exceeds 10^{15} $L_\odot$, the three-image configuration is more appealing. However, the core ra-

Restframe wavelength   (μm)

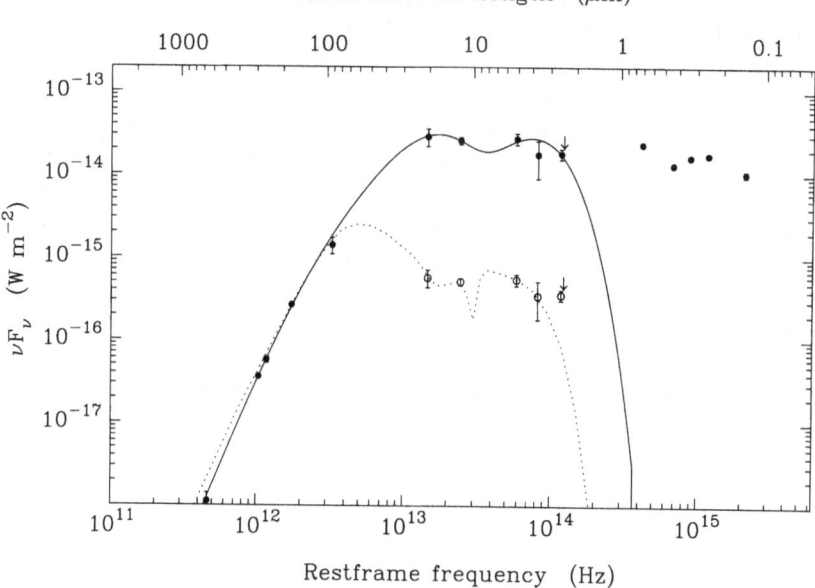

Fig. 4.12. The spectral energy distribution of APM 08279+5255 fitted by two isothermal greybody models. The full-drawn line corresponds to the sum of a 'cool' component ($T_{dust} = 200$ K) and a hot component ($T_{dust} = 910$ K). The dust emission becomes optically thick at $\lambda < 200\mu$m. The data points (filled circles) are from [70,85,89,43,47] and the Faint Source IRAS catalog (NED). The dotted line is a starburst model of Rowan-Robinson & Efstathiou [121], which has been arbitrarily fitted to the long wavelength part of the SED. The data points marked by open circles are the equivalent observed data points but with their values reduced by a factor 50 in order to fit on the starburst model.

dius is large, $0\rlap{.}''21$, almost as large as the Einstein radius, $0\rlap{.}''29$. If the lens is at a redshift $z_d \approx 1.2$, the core radius corresponds to ~ 1.2 kpc in the lens. This is much larger than most measured core radii. In a survey of 42 giant elliptical galaxies it was found that the objects which can be resolved have a median core radius of $225h^{-1}$ pc [84]. In the case of a three-image model, the potential is almost circular with $\epsilon = 0.012$, while the two-image model gives $\epsilon = 0.083$ [47]. The difference in ellipticity corresponds to a difference in the size of the caustic structure, which in turn influences the effects of differential magnification. In the three-image model, the caustic structure is approximately 45 pc in extent in the source plane, while the two-image model has a caustic structure almost 5 times larger.

The effects on the lensing behavior for a source with a finite extent was explored in [47]. A source with an extent exceeding ~ 500 pc resulted in a filled disk. A more detailed model was done [116] where an assumed source temperature distribution was used in order to derive the resulting spectral energy distribution. Using the model parameters of [47], where the source is located between the radial and tangential caustic for the three-image model, the hot dust is expected to be moderately enhanced by the outer magnification plateau (cf. Fig. 4.10). In the two-image scenario, the QSO is again located outside the tangential caustic. In this case, however, the radial caustic is lacking due to the singular potential. The latter scenario can produce a modest negative distortion of the SED (as seen in the top left panel of Fig. 4.13). In the three-image scenario, however, the effect on the SED is more dramatic and represents a positive distortion, i.e. the mid-IR part of the SED is enhanced relative to the long wavelength part (bottom left panel of Fig. 4.13). The magnitude of the distortion is quite large, its details depending on the extent of the dust region. For a dust distribution with a radius of 650 pc, the differential magnification can enhance the intrinsic flux at restframe mid-IR wavelength with a factor $\sim$ 10. A smaller extent of the dust results in a smaller enhancement factor. The dust region (for a radius of 300 pc) and the caustic structure are seen in the lower right panel of Fig. 4.13.

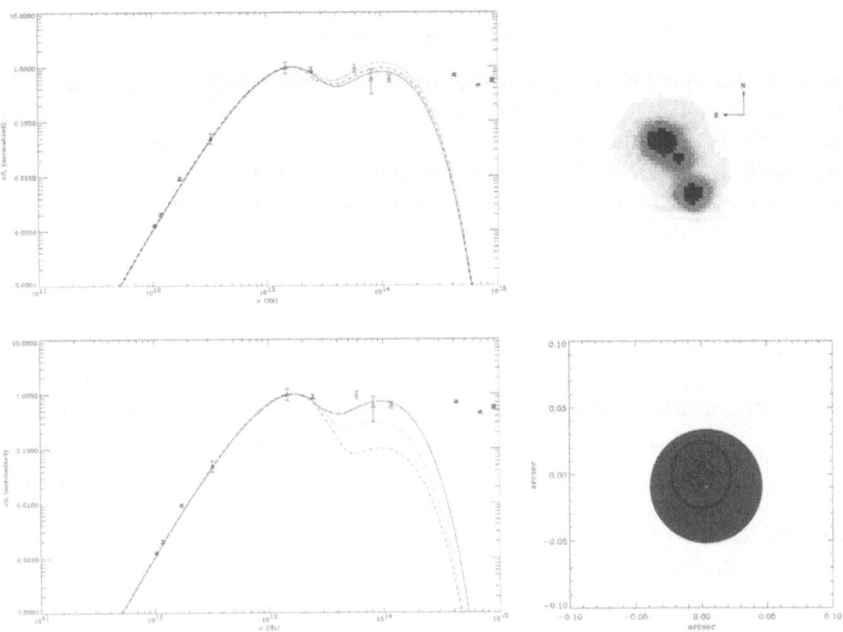

Fig. 4.13. The 2-image and 3-image lens model solutions for APM 0927+5255 and their effect on the spectral energy distribution. In the 2-image solution (top left), the intrinsic mid-IR part of the SED is slightly depressed, while in the 3-image solution (bottom left) the intrinsic mid-IR is strongly enhanced (representing a positive distortion, see Sect. 4.4.5). The intrinsic SED is represented by dashed and dotted lines, for a dust distribution with a radius of 600 pc and 300 pc, respectively. The full drawn line is the observed SED. In the top right is APM 08279+5255 seen with the HST WFPC2 camera and lower right is the dust model and its location used in the 3-image solution. From K. Pontoppidan's Master Thesis, Copenhagen University [116].

The three-image model of APM 08279+5255 is more likely to be correct than the two-image model since it produces a magnification which corresponds to a source with a bolometric luminosity which is large, but not extreme. The three-image model also means that for realistic dust distribution, the restframe mid-IR is strongly enhanced relative to the longer wavelength part of the SED. The intrinsic SED of APM 08279+5255 resembles that of less extreme dusty QSO spectra (cf. [30]). In fact, the shape of the intrinsic SED of APM 08279+5255 now resembles that of pure starburst models, except that the mid-IR is still enhanced by a factor $5 - 10$, marking the influence of the AGN. This shows that the effects of differential magnification must be considered before applying radiation transfer models to gravitationally lensed dusty sources.

In order to model the resolved and extended low-J CO emission, the effects of a highly elliptical lensing potential has been explored [91]. The lensing galaxy is here assumed to be an edge-on spiral. The result is a good fit with the extended low-excitation CO emission, while the point sources from the background QSO, although imaged into three components, have widely different magnification ratios compared to the observed values. This may not be of great importance if microlensing affects the optical photometric results (e.g. [90]).

4.5.2   The Cloverleaf: Another Case of Differential Magnification

The Cloverleaf is the gravitationally lensed image of a BAL quasar at $z = 2.558$, H1413+117, showing four quasar-images (hereafter called spots) with angular separation from $0\overset{''}{.}77$ to $1\overset{''}{.}36$. Since its discovery [96], the Cloverleaf has been imaged with ground based telescopes in numerous bands up to I and with HST/WFPC2 in the UV, optical and near-IR [146,77,78].

The Lensing System.   After the early lens model of Kayser et al. [76], these new data sets have been used to derive an improved model of the lensing system [77,78] which now includes:

1. A cluster of galaxies with derived photometric redshifts in the range 0.8 to 1.0, which contributes to the magnification.
2. A lensing galaxy close to the line of sight to the quasar, which determines the geometry of the image (four main spots) and carries the largest share of the magnification. The redshift of the lensing galaxy has been tentatively measured with VLT/ISAAC at a value of 0.9 (Faure et al. in prep).

In the following, we use this new model for the lensing system, which is essentially constrained by the HST data. Further details can be found in [77,78].

The IRAM Millimeter Data Sets.   After its discovery in the CO(3-2) line emission with the IRAM Pico Veleta dish [12], the Cloverleaf has been observed in the millimeter range by various teams and instruments ([160] with BIMA, [164] with OVRO). Yet, the best data sets collected to date on this object are from the IRAM Pico Veleta dish and Plateau de Bure interferometer.

A total of six millimeter transitions have been reported from observations with the IRAM Pico Veleta dish: CO(3-2), CO(4-3), CO(5-4), CO(7-6), CI(^{3}P$_1$-^{3}P$_0$) and HCN(4-3) [13]. Detailed non-LTE modeling of the CO line strengths by these authors indicates that the molecular gas is warm (T larger than 100 K), dense (n(H$_2$) density larger than 3×10^3 cm^{-3}) and not very optically thick ($\tau_{CO} < 3$). These results suggest that the molecular material is close to a powerful heating source and might therefore be related to the environment of the central engine in the quasar. They also prompt us for not using the conventional conversion factor CO to H$_2$ which is derived for molecular clouds in the disk of our Galaxy.

Thanks to the strength of the CO(7-6) transition, a high resolution (0$.''$5) map was obtained with the IRAM Plateau de Bure interferometer. A first CO(7-6) interferometric data set [3] has later been complemented with observations at intermediate baselines [77]. The combined data has lead to the CLEANed map restored with an 0$.''$5 circular beam, shown in Fig. 4.14 . In order to search for a velocity gradient, we have derived the spatially integrated line profile, following the procedure described in [3]. The CO(7-6) line profile (Fig. 4.15) shows a marked asymmetry with a steep rise and excess of emission (with respect to a standard Gaussian) on its blue side and a slower decrease on its red side. Excluding the central velocity channel (so that the split in velocity is symmetric), we have built the blue (-225, -25 km/s) and the red (+25, +225 km/s) maps displayed in Fig. 4.14c,d respectively. The difference between the red and blue CLEANed maps (Fig. 4.14b) establishes firmly the presence of a velocity gradient at the 8σ level. Measurements of the characteristics of the spots from the CO(7-6) image have been performed (spot flux ratios, sizes and orientations) through a fitting procedure in the visibility domain, as explained in [3]. Final parameters are provided in Table 1 of [77] where the spot sizes are intrinsic to the image, i.e. deconvolved by the interferometer beam: spots A, B and C definitely appear elongated.

Comparing Images in the UV and the Millimeter Range. Images in the UV/optical correspond in the quasar restframe to the emission from the accretion disk surrounding the quasar central engine. This latter source is expected to be point-like. The four spots on the HST images do indeed have a stellar-like appearance, being circular with a FWHM of about 0$.''$068 [77]. The absolute photometry and relative intensity ratios of the four spots (Table 4 in [77]) have been computed using the Sextractor software [14]. The large variation of the intensity ratios in U, compared to V, R and I bands can probably be explained by absorption along the line of sight by intervening galaxies. Alternatively, this effect can be ascribed to dust extinction at the redshift of the quasar, using an SMC-like dust extinction law [146]. The presence of such an absorbing medium in the close environment of the central engine could be put in relation with the BAL appearance of this quasar.

The millimeter CO lines are expected to arise from an extended structure, the so-called dusty/molecular torus, with an intrinsic radius of a few 10 pc to a few 100 pc (according to models). In such a configuration, different parts of

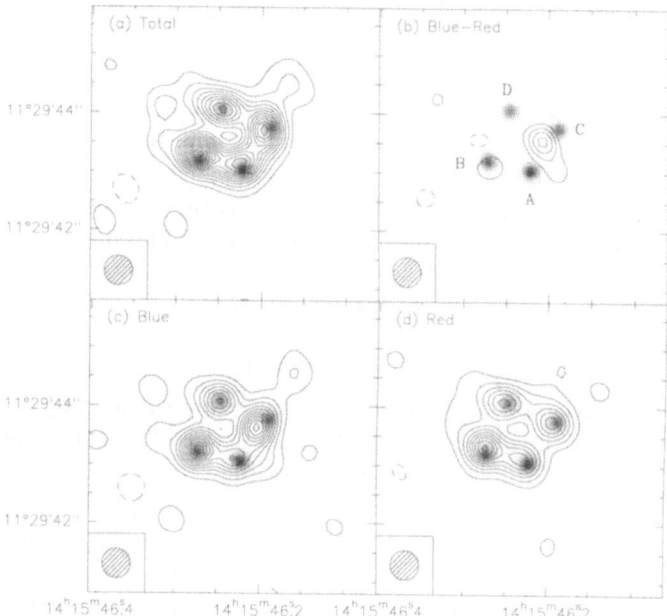

Fig. 4.14. Image of the Cloverleaf obtained with the IRAM Plateau de Bure interferometer [77]. **a)** is the total CLEANed image. **c)** and **d)** are the blue and red part of the total emission profile, while **b)** shows the difference between the CLEANed blue and red images. A velocity gradient in the underlying source can be inferred from the residual. The data has been restored with a circular beam of angular resolution $0\overset{''}{.}5$.

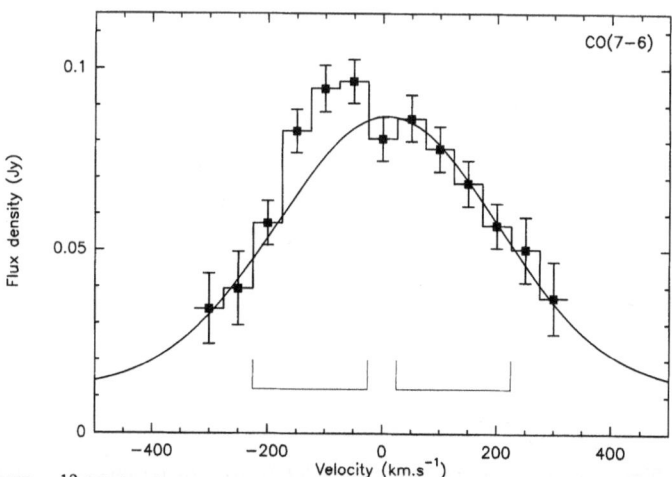

Fig. 4.15. The ^{12}CO(7-6) spectrum from the $z = 2.56$ Cloverleaf quasar obtained with the IRAM Plateau de Bure interferometer [77]. The thin line represents a best fit Gaussian profile. Notice the asymmetric line profile with excess emission at the blue part of the spectrum.

the extended torus will be positioned differently with respect to the caustic (the curve which represents in the source plane the signature of the lensing system). As the image properties and the amplification factor in particular, are ruled by the relative positioning of the source/caustic, the four spots on the CO(7-6) image, each corresponding to the extended torus, will be distorted with respect to the four spots on the HST image (corresponding each to a point-like source). This features what is called 'differential magnification effects' (see Sect. 4.4.5). From the blue bump appearing on the CO(7-6) line profile (Fig. 4.15), we clearly see that the blue-shifted part of the CO line arises from a region of the molecular torus which is positioned closer to the caustic than the region emitting the red-shifted side of the CO line. In this way, we are able to recover detailed structural and kinematical information about the molecular torus in the quasar.

In order to derive precisely the shear induced by the lensing system on the extended source in the quasar, it is imperative to register with a high accuracy the Cloverleaf image in a waveband corresponding to a point-like source in the quasar (accretion disk: UV restframe, that is an R band image for example) and in a waveband corresponding to an extended source in the quasar (molecular torus: CO(7-6) line). The high precision required, better than $0\rlap{.}''2$, was achieved using a combination of the HST data and of CFHT data acquired over a larger field of view under extremely good image quality [77].

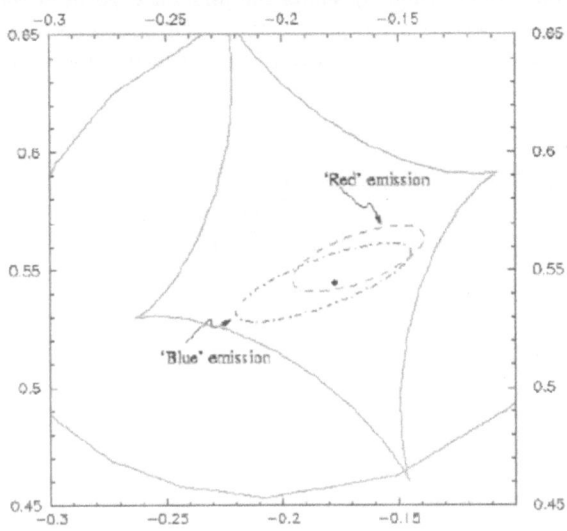

Fig. 4.16. The caustic structure and the CO source distribution in the Cloverleaf quasar [77]. The central ellipses represent the CO source distribution for the red and blue part of the emission profile, respectively. The dot in the center corresponds to the quasar UV point source. The scale on the axis is arcseconds. The result has been obtained by combining an HST optical image with the interferometric CO data.

Derived Properties of the Molecular Torus in the Cloverleaf BAL Quasar at $z = 2.558$. We have used the model of the lensing system as constrained by the HST data and presented above. The total amplification factor for the CO emission is found to be 30. Correcting for this amplification factor, one derives a molecular gas mass $M(H_2) = 2 \times 10^9$ $M_\odot$ and an atomic hydrogen gas mass $M(HI)= 2 \times 10^9$ $M_\odot$ [13,77].

We have derived the properties of the molecular torus in the quasar using the CO(7-6) maps: firstly using the total line flux and secondly, using separately each of the blue and red halves of the CO(7-6) line [77]. We find a typical size for the molecular torus of 150 pc (assuming $H_0 = 50$ km s^{-1}, $\Omega_m = 1$ and $\Omega_\Lambda = 0$). When we treat separately the maps corresponding to the blue-half line and red-half line, we find that the quasar point-like UV source is almost exactly centered between the region emitting the blue-half of the CO line and the region emitting the red-half of the line (Fig. 4.16). This is reminiscent of a disk- or ring-like structure orbiting the quasar at about 75 pc and with a Keplerian velocity of 100 km s^{-1} (assuming a disk orientation perpendicular to the plane of the sky). The resulting central dynamical mass would be about 8×10^8 $M_\odot$. This value is in good agreement with the estimate of the molecular gas mass made above from the total CO line flux, provided uncertainties in the inclination of the molecular torus and in the conversion factor from I(CO) to N(H$_2$).

In conclusion, we can regard the case of the Cloverleaf as a first and enlightening example of what will become routine when ALMA becomes available. Indeed, exploiting differential magnification effects is an extremely promising technique. The effective angular resolution on the CO source in the quasar at $z = 2.558$, using this procedure, is $\sim 0''.03$, or about 17 times smaller than the synthesized beam of the IRAM observations! And the amplification factor in this case is around 30!

4.5.3   PKS 1830-211: Time Delay and the Hubble Constant

Observations and applications of differential time delays in gravitational lenses are discussed in detail Chap. 1. Here some results which have implications for the millimetric part of the electromagnetic spectrum will be presented. More specifically, we will discuss a derivation of the differential time delay in the gravitational lens PKS 1830-211 obtained from the saturated molecular absorption line of HCO$^+$(2-1).

The gravitational lens system PKS 1830-211 has been described in some detail above (Sect. 4.3.3). The background quasar is variable at radio wavelengths, with an amplitude which increases at shorter wavelengths. This is due to the fact that the core, where the variability occurs, is a flat-spectrum source while the jet, which has a more or less constant flux, has a steep radio spectrum. It is presently unknown if PKS 1830-211 is variable at infrared/optical wavelengths, although this is likely to be the case.

Time delay measurements of the PKS 1830-211 system has also been done using long wavelength radio continuum [147,94]. In one case a single dish telescope was used and the two main lens components were not resolved [147]. The

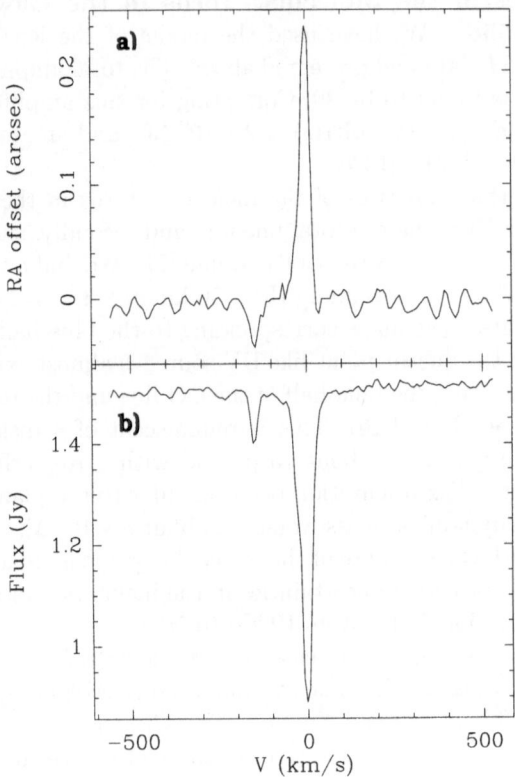

Fig. 4.17. Bottom: Spectrum of HCO$^+$(2-1) at $z = 0.886$ towards PKS 1830-211 obtained with the IRAM interferometer. The main absorption line is seen around zero velocity. A secondary, weaker absorption line of HCO$^+$(2-1) is seen at a velocity of -147 km/s relative to the main line. **Top :** The right ascension shift of the phase center of the continuum emission as a function of velocity. A negative shift means that the phase center moves towards the NE component, while a positive shift indicates a shift towards the SW component. Comparison with the absorption spectra shows that the main absorption component covers the SW source, while the weaker secondary absorption covers the NE source. (From [154]).

analysis had to be based on a compound light curve and the derived time delay of 44 ± 9 days should therefore be regarded as tentative. In the other case, the ATCA interferometer was used, with an angular resolution that did not fully resolve the NE and SW components [94]. Instead a model fitting procedure was used in order to obtain two separate light curves over an 18 months period. The resulting differential time delay is 26^{+4}_{-5} days. Although the analysis is model dependent, this result represents a considerable improvement in the Δt estimate.

Time Delay Measurements Using Molecular Absorption Lines. As discussed in Sect. 4.3.3, the lens in the PKS 1830-211 system was first detected through molecular absorption lines at a redshift $z_d = 0.88582$ [151]. More than

16 different molecular species in 29 different transitions have so far been detected at millimeter wavelengths [151,154,56]. Two additional molecular species in three different transitions have been observed at cm wavelengths [97].

The millimeter transitions include three different isotopic variants: $H^{13}CO^+$, $HC^{18}O^+$ and $H^{13}CN$. The mere detection of these lines shows that the main isotopic transitions of these molecules must be highly saturated. Despite this the absorption lines do not reach zero intensity (Fig. 4.17b). This can only be reconciled with an optical thick obscuration that do not completely cover the background continuum emission. In fact, from the ratio of the total continuum and the depth of saturated molecular absorption lines (such as $HCO^+(2-1)$, $HCN(2-1)$, etc.), it was concluded that only the SW lens component is obscured by molecular gas and that the covering factor of this particular image is unity or close to unity [151]. A secondary weaker molecular absorption has now been found towards the NE component as well, separated in velocity by $-147 \, km \, s^{-1}$ [154].

Imaging of the $HCO^+(2-1)$ absorption line with the IRAM millimeter wave interferometer did not directly resolve the NE and SW components [154]. This is due to the low declination of the source relative to the latitude of the interferometer, creating a synthesized beam elongated in approximately the same direction as the image separation. The continuum, however, is strong enough to allow self-calibration, making it possible to accurately track the phase center. The best angular resolution is achieved in right ascension ($\sim 0''\!.1$) with a factor ~ 2 worse resolution in declination due to an elongated synthesized beam. At frequencies outside the absorption line, the phase center should fall on a line in between the NE and SW components. Assuming that the flux ratio NE/SW is similar to that derived for longer wavelengths ($\sim 1.3 - 1.4$), the phase center should move towards positive RA at frequencies where the absorption occurs. If the covering factor is unity, $\Delta\alpha$ should be $\sim +0''\!.25$. This is exactly the amount of shift observed for the saturated $HCO^+(2-1)$ line (Fig. 4.17a). A similar shift in the declination of the phase center can also be seen and concurs with these results [154]. This result has also been confirmed through BIMA observations where the two continuum components have been separated [140].

Due to this fortunate configuration of obscuring molecular gas in the lensing galaxy, the flux contributions from the NE and SW cores can easily be estimated using molecular absorption lines and a single dish telescope with low angular resolution. The 15m SEST telescope, which is used for the time delay monitoring presented here, has a HPBW of $\sim 50''$ at the observed frequency of the $HCO^+(2-1)$ transition, much larger than the image separation of $0''\!.97$. Since molecular gas covers only the SW component, as shown by the interferometric data, and the line opacity is $\gg 1$ as seen from the rare isotopic lines, the depth of the absorption line corresponds to the flux from the SW component only. The total continuum away from the absorption line corresponds to the sum of fluxes from the SW and NE components (Fig. 4.18).

Monitoring of $HCO^+(2-1)$. Monitoring of the $HCO^+(2-1)$ absorption and the total continuum flux has been going on at the 15m SEST telescope since

The gravitational lens PKS1830-211 and the z=0.89 absorption line system

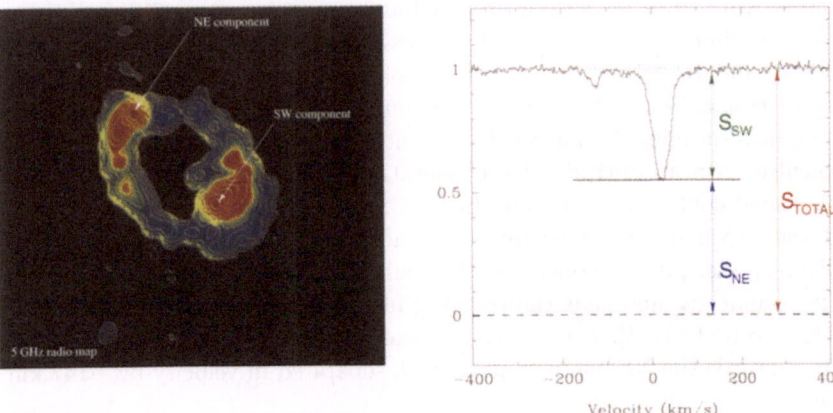

Fig. 4.18. Illustration of how the individual light curves of the NE and SW components can be derived from the single dish millimetric observations. A 5 GHz radio image of PKS 1830-211 [139] shows the two cores and the extended jet emission. At millimeter wavelengths only the cores contributes to the continuum emission. The right hand panel shows the saturated HCO^+ (2-1) spectrum. Since the molecular gas only covers the SW component and the line opacity is $\gg 1$, the depth of the absorption line corresponds to the flux from the SW component only. The total continuum away from the absorption line corresponds to the sum of the fluxes from the SW and NE components (see Color Plate).

April 1996. Data can only be obtained between February and November due to Sun constraints (PKS 1830-211 comes within 3° from the Sun). The light curves are shown in Fig. 4.19. Since only the total continuum (at the top in the figure) and the depth of the absorption line (at the bottom in the figure) are measured, the flux from the NE component (middle) is derived as the difference $T_{NE} = T_{tot} - T_{abs}$, and therefore has a somewhat higher uncertainty. During the 1996-2000 campaigns a total of 144 usable observations have been obtained. The background quasar had a large outburst during 1998. The outburst shows a single peak in the total continuum, putting an upper limit to the differential time delay between the two cores (no double peak structure). Separating the light curves for the two cores, however, one can clearly see a delayed response of the SW image relative to the NE (Fig. 4.19). This is seen even more clearly when plotting the ratios of the NE and SW fluxes (Fig. 4.20). This ratio shows the relative magnification of the two cores and should be constant in the absence of a time delay. The decrease in the flux ratio during the 1998 outburst is a clear indication of the time delay between the NE and SW cores.

Monitoring Results. The light curves shown in Figs. 4.19 and 4.20 have been analyzed using several different techniques. The problem is straightforward: correlate two unevenly sampled time series. However, the analysis is complicated

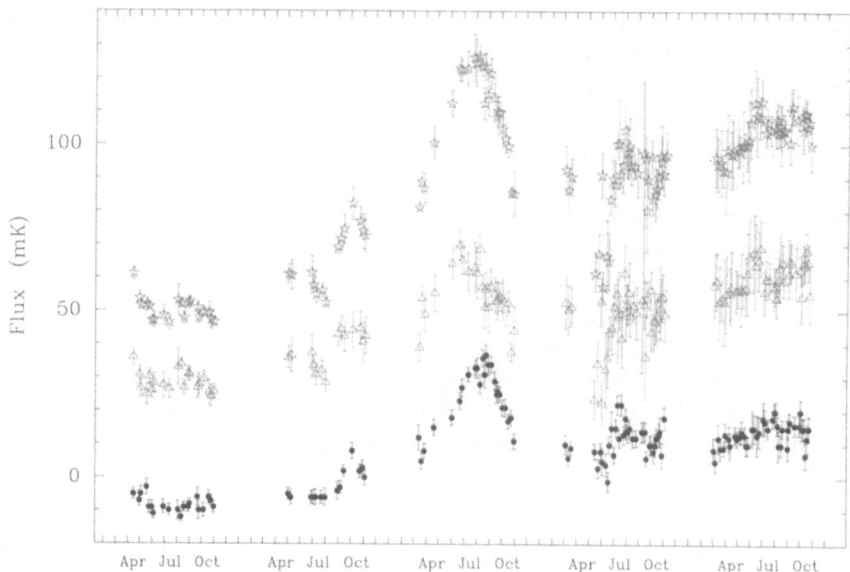

Fig. 4.19. Results from 5 years of monitoring of the HCO^+(2-1) absorption at $z_d =$ 0.886 towards PKS 1830-211. The top curve shows the measured total continuum flux away from the absorption line. Notice the large outburst during 1998. The bottom curve shows the depth of the HCO^+(2-1) line (shifted by -30 mK). This corresponds to the flux from the SW component. The middle curve shows the flux derived for the NE component.

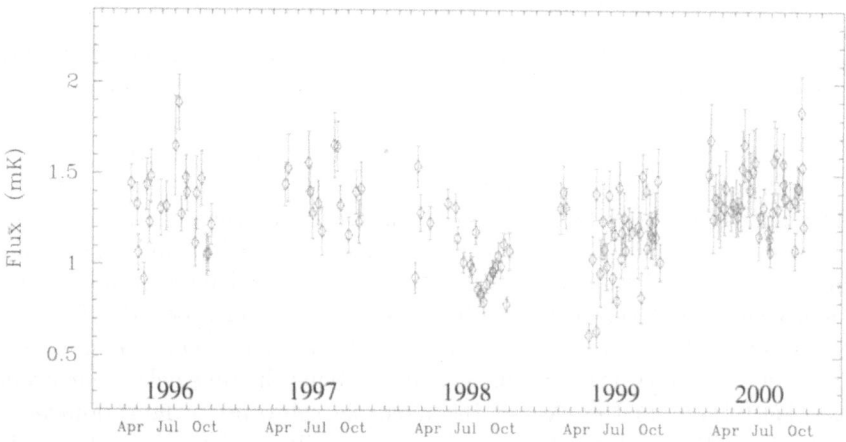

Fig. 4.20. Same as Fig. 4.19 but showing the flux ratio of the NE and SW components.

by the fact that since the two time series are really copies of each other but shifted in time they are effectively sampled at different epochs.

There are two main types of methods used for obtaining the time delay. One is to interpolate unobserved data points and then apply standard techniques for cross correlation. The other is to use the unevenly sampled time series and try to correlate neighboring points as good as possible. The former method is superior if the sampling rate is high, but this is usually not the case for astronomical data. The latter method has less precision but is the least unbiased way of obtaining the time delay. Both types of methods have been used for the molecular absorption line data on PKS 1830-211.

Data Analysis. The observables at each epoch t_i are the total continuum flux $S_0(t_i)$ and the depth of the absorption line, corresponding to the flux from the SW component $S_{SW}(t_i)$. The total continuum is the sum of the two image components,

$$S_0(t_i) = S_1(t_i) + S_2(t_i) \ ,$$

where subscript 1 refers to the NE component and 2 to the SW component. We know that the S_1 and S_2 fluxes are related as

$$S_1(t_i) = \mu \, S_2(t_i + \Delta t) \ ,$$

where μ is the magnification ratio and Δt the differential time delay between the two cores. Hence,

$$S_0(t_i) = \mu \, S_2(t_i + \Delta t) + S_2(t_i) \ .$$

By shifting the observed S_2 value, multiplying it with a magnification ratio and adding the observed unshifted and unmagnified value we should recover the observed total flux at time t_i. Since the observations consist of an unevenly sampled time series, with significant amount of noise, finding the true Δt and μ is a non-trivial exercise.

The analysis of the light curves has been done using three methods. Two of them involves interpolating between the observed data points and construction of an evenly sampled time series. Due to the rather long interruptions due to the Sun avoidance, interpolation only extends over periods between February and November. An elaborate interpolation scheme of unevenly sampled data has been developed with the specific goal of resolving the time delay controversy of 0957+561 [117]. In the case of the molecular absorption data, however, a smoothing function with an effective resolution similar to the average data point separation at each epoch was applied. Using the smoothed time series, data points in between observed epochs were linearly interpolated. The smoothing dampens the worst fluctuations while retaining the small scale structure in the time series. It also allows an easy assessment of the relative weights of observed and interpolated data. A complication, however, is that the data points are no longer completely independent. This is of some concern when deriving the reduced χ^2 values.

Smoothed and interpolated data

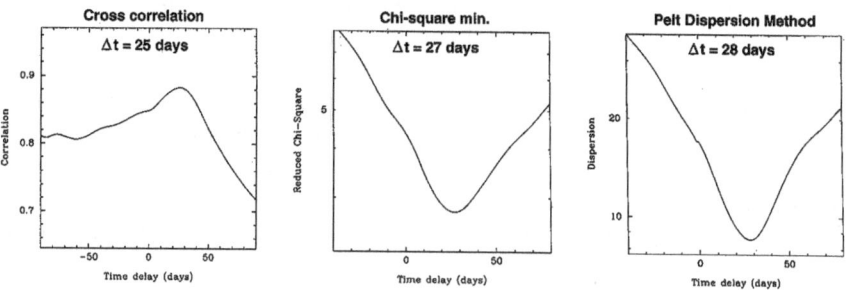

Fig. 4.21. The time delay for PKS 1830-211 derived from molecular absorption lines and using cross-correlation, χ^2 minimization and the Pelt Minimum Dispersion method. All three methods in this example use the interpolated data set.

χ^2 **Minimization** : Minimization was done using both the time delay Δt and the magnification ratio μ as well as keeping the magnification ratio fixed or time dependent. The latter is due to a surprising realization that the magnification ratio might be variable, albeit on a much longer time scale than the time delay (cf. Fig. 4.20). Using a parameterized $\mu(t_i)$ means that the solution becomes cumbersome and slow. Instead we smoothed the flux ratio and fitted a third order polynomial. This parameterization of $\mu(t_i)$ was used when solving for Δt. The result, together with results from the other analysis methods, is shown in Fig. 4.21 and gives a $\Delta t = 27$ days.

Cross Correlation: Edelson and Krolik [46] developed a discrete cross correlation method specifically aimed for reverberation mapping of AGNs that can be used for time delays in gravitational lensing. The method optimizes the binning of data points rather than the interpolation, as in the method of Press et al. [117]. The method requires a fairly well sampled data set to start with in order to retain a sufficiently good temporal resolution. The sampling rate for molecular absorption line data in PKS 1830-211 is not dense enough to use the Edelson & Krolik method. Instead cross correlation was done on the same smoothed and interpolated data set as the χ^2 minimization. The cross correlation coefficient is defined as $r_{ab} = s_{ab}^2/(s_a \, s_b)$, where the covariances s_a and s_b are defined in the usual manner (cf. [15]). As with the χ^2 minimization, the data points are not entirely independent due to the smoothing and interpolation and the variances are only approximately true. The result gives $\Delta t = 25$ days, with a rather broad maximum for the cross correlation coefficient (Fig. 4.21).

Minimum Dispersion (The Pelt Method): A simple and robust technique for analyzing unevenly sampled time series was presented by Pelt et al. [110,111]. They successfully applied it to the lens system 0957+561. The strength of the method is that interpolation or smoothing are not needed, leaving the errors for each data point independent. The method is a form of cross correlation where a given data point is correlated with a data point which is temporally its closest neighbor. The method is illustrated in Fig. 4.22, where the round and square markers in the two top rows represent the two photometric data sets obtained from a two-component gravitational lens. When correlating the time series, one of them is shifted in time, as the square markers in the middle rows. Projecting both the unshifted (round) and the shifted (square) time series to a common array (bottom row), correlation is done between those data points which are from different time series and closest to each other. These points are connected by arcs in the figure. It is easy to include the effects of different magnifications for the lensed components as well as time delays in systems with two or more lenses [112].

The results are undeniably noisier than for the interpolated data sets. This can be seen in Fig. 4.23, where the Pelt dispersion method has been applied to both raw and interpolated data. The best fit is for a time delay $\Delta t = 28$ days, with the NE component leading.

Error Analysis: The errors associated with the light curves are a combination of noise in the data points and systematic errors. The latter can originate in the instrument, in the modeling necessary for separation of the lensed components (as in the case of long wavelength radio observations), assumptions made about the lensing system, etc. When interpreting the time delay in terms of a Hubble constant, the largest systematic error comes from modeling of the gravitational potential (see Chap. 1). Noise in the data comes from imprecise measurements but can also originate in secondary variability such as microlensing and interstellar scintillation. The latter is applicable at long radio wavelengths. Microlensing may be of importance even for gravitational lenses observed at radio wavelengths (cf. [81]).

In order to assess the significance of correlations found in the light curve of gravitationally lensed images it is customary to derive the confidence limits through Monte Carlo simulations and bootstrap techniques (cf. [51]). The results often show non-Gaussian distributions and confidence levels are set by finding the range of delays and magnification ratios inside which a given amount (say 95%) of the simulations lies. This gives a better estimate of the true confidence level than simply fitting a Gaussian to the distribution. Doing this for the molecular absorption line data in PKS 1830-211 gives a time delay of $\Delta t = 28^{+4}_{-5}$ days, with the NE component leading.

In Fig. 4.24 the light curve of the SW component in PKS 1830-211 has been shifted by -28 days and multiplied by a magnification ratio μ. In the upper panel a constant ratio of $\mu = 1.3$ was used, while in the lower panel a time dependent magnification ratio was used. The use of different parameterizations of the magnification ratios do not change the derived time delay, but the time dependent form provides a better fit of the two light curves. The reason for the slow change in magnification ratio is presently unclear. It may have implications for the use of molecular absorption lines as a probe of the time delay, but since the time scale for the change of the magnification appears to be much longer than the time delay, it is likely to be of small importance when correlating light curves for each period (i.e. 9 months).

Original observation

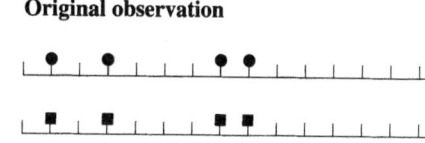

Shifted version

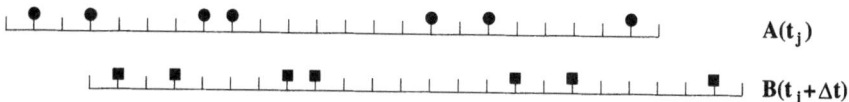

Resulting array

Fig. 4.22. Illustration of the Pelt Minimum Dispersion method used to determine time delays in gravitational lenses. This method was one of the methods used for deriving the time delay in PKS 1830-211 from molecular absorption lines. This particular case shows a two-image lens (A and B), where the respective light curves are sampled at irregular intervals. In the middle section the B light curve is shifted by Δt. By projecting the resulting data points to a common array (C), nearest neighbors of different light curves are correlated (arcs). A weight, depending on the time difference between the points used in the correlation, can be applied.

4.6 Lens Models for PKS 1830-211

In order to use the differential time delay to derive a value for the Hubble constant, a lens model has to be fitted to the observed data. This is not a trivial exercise in most cases and this is particularly true for PKS 1830-211. Due to its location at Galactic longitude $l = 12.2°$ and latitude $b = -5.7°$, PKS 1830-211 suffers considerable Galactic extinction. In addition, the molecular gas seen in absorption towards the SW component contributes significant obscuration for at least this image. Early attempts to identify the radio source PKS 1830-211 with an optical counterpart were all unsuccessful [139,39]. It was only with the advent of sensitive infrared imaging and spectroscopic capabilities that progress could be made. The NE image was positively identified using K-band imaging at Keck and the ESO NTT [36]. While the redshift of the lens, $z_d = 0.886$, had been derived using molecular absorption lines [151], the redshift of the source was obtained from near-infrared spectroscopy [92]. The redshift was found to be $z_s = 2.507$. Imaging with the HST WFPC2 and NICMOS allowed identification of both the NE and SW image [87]. In addition, an object which might be the lensing galaxy was detected (designated as G). Its exact center position remains

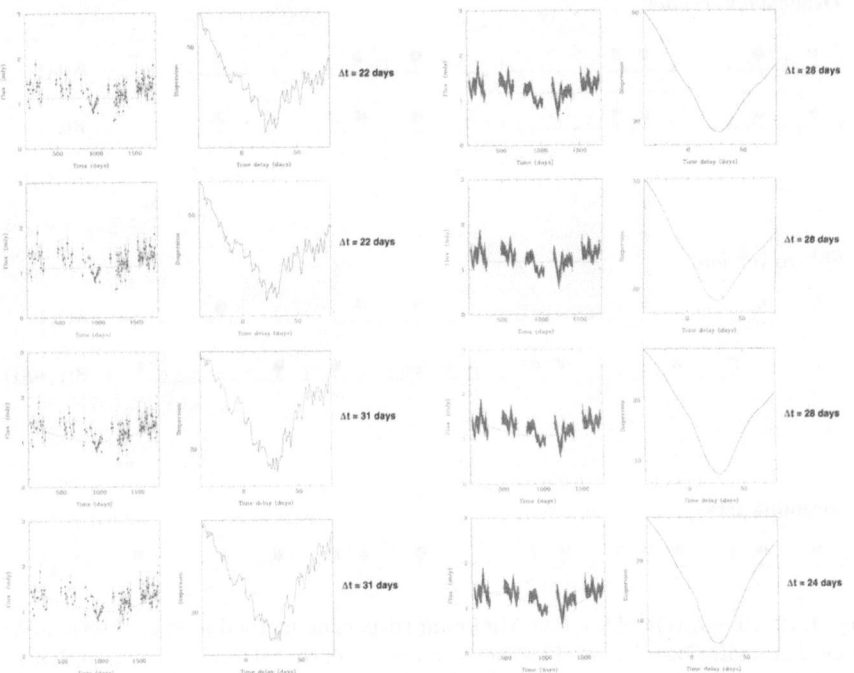

Fig. 4.23. Results for the time delay of molecular absorption lines in the PKS 1830-211 gravitational lens system using the Pelt Minimum Dispersion method. Both raw and interpolated data sets are shown (left and right, respectively). Also shown are the results for different treatments of the magnification ratio. In the top two rows, the magnification ratio is included in the fit, while in the two bottom rows the magnification ratio is predetermined by fitting either a second or third degree polynomial to the observed magnification ratio.

uncertain due to the presence of a point source ~ 190 mas away. The nature of the point source remains unknown but could possibly be a Galactic star and thus of no importance for the lens model. In a recent paper by Courbin et al. [37] combined images from the HST and Gemini-North telescopes show what might actually be the lensing galaxy at $z_d = 0.889$. The lens has two spiral arms, as expected from the molecular absorption data. One spiral arm crosses the SW image of the QSO. The center of the spiral is, however, significantly offset from the line joining the NE and SW images. Based on symmetry arguments, the center of the lensing galaxy is believed to be in the proximity of this line joining the images (Fig. 4.25).

All the necessary ingredients for a detailed lens model are thus in place, except for two remaining uncertainties: the exact position of the lensing galaxy and the possible double-lens nature of the system (cf. Sect. 4.3.3).

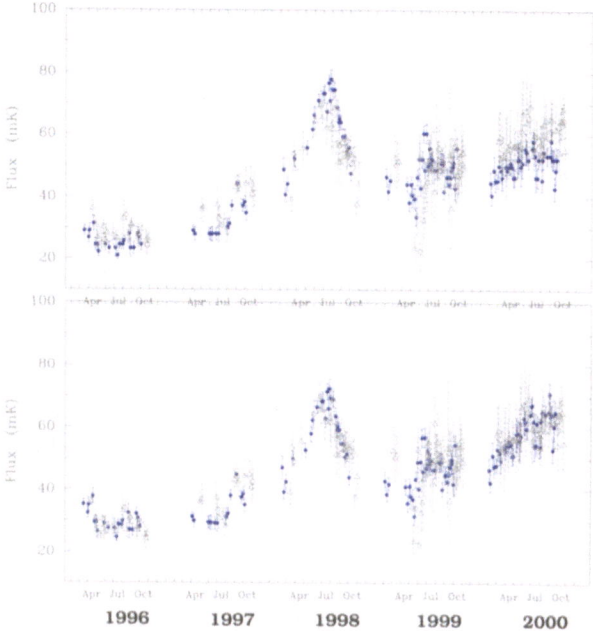

Fig. 4.24. The light curve of the SW component in PKS 1830-211 (black) shifted by $\Delta t = -28$ days and multiplied by a magnification ratio μ. The light curve for the NE component is shown in grey. In the upper panel a constant magnification ratio of $\mu = 1.3$ was used, while in the lower panel a time dependent magnification ratio was used (cf. Fig. 4.23) (see Color Plate).

4.6.1   Early Models

There have been several attempts to model the lens system PKS 1830-211. The system was first detected at radio wavelengths, where it is a prominent southern radio source. The morphology of the system was found to be that of a double, while the radio spectrum is typical for a compact flat-spectrum source. This led [120] to first suggest that PKS 1830-211 is a gravitationally lensed system. Based only on the radio images and their polarization properties, obtained with the Very Large Array (VLA) at 5 and 15 GHz, [139] constructed a lens model which is not much different from later ones based on more detailed data. In order to reconstruct the extended radio structure, [139] modeled the source as an one-sided core-jet structure. To get lensed images with a morphology similar to the observed one, they also had to include a 'knot' in the jet. Based on their lens model Subrahmanyan et al. predicted a time delay of $27\,h_{100}^{-1}$ days (using the now known redshift of the source and the lens).

Nair et al. [102] modeled the PKS 1830-211 system using improved radio interferometry data (cf. [75]). The method was similar to that of [139] in that the source structure was built up in a piecemeal manner in order to fit various observed features. With this type of method one can emphasize the influence of small and weak features which may carry a small weight in an inversion scheme

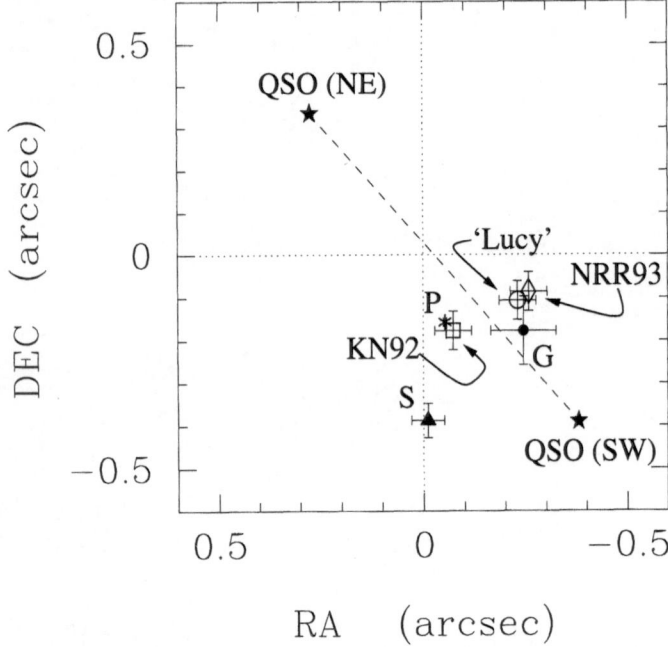

Fig. 4.25. Illustration of the positions of various components in the PKS 1830-211 gravitational lens system. The two cores are marked by QSO (NE) and QSO (SW). The putative location of the lens galaxy, derived by Lehár et al. [87], is marked by 'G' (filled circle) and a point source of unknown origin is marked by 'P' (star). A new possible lens location (from [37]) is marked by 'S' (filled triangle). The location of the center of the lens derived from the lens model here, using the Lucy rectification scheme, is marked by 'Lucy' (open circle), the lens center derived by Nair et al. [102] by 'NRR93' (open diamond) and the lens center derived by Kochanek & Narayan [79] by 'KN92' (open rectangle). The dashed line joining the two QSO images is only to guide the eye.

based on χ^2 minimization of model−observed results, but may nevertheless carry important information on the lensing scenario. In the PKS 1830-211 system such a weak radio feature, labeled E, was used by Nair et al. to constrain the lens model. This feature might be a third demagnified image of the core. However, since the flux of the E component is less than one percent of the peak value, its significance in terms of flux is small unless the dynamic range of the interferometry maps is very good. Nair et al. found that an elliptical potential with the radio core located close to the inner edge of the radial caustic gave a good fit to the observed morphology. As in the previous model by [139], it was necessary to include a 'knot' feature in the source distribution. The jet needed to be bent and cross the tangential caustic. The model gives a good fit to the observed system and places the lens galaxy close to, but not coinciding with, the possible lens position observed by [87]. The estimated time delay, using the known source and lens redshifts, is $17\,h_{100}^{-1}$ days.

A difficulty with extended lensed images is that any inversion must solve simultaneously for both the lens configuration and the source structure. This type of inversion problem can be seen as

$$I_{obs}(\zeta) = \int \psi(\zeta') \, K(\zeta - \zeta') \, d\zeta' \ , \tag{4.7}$$

where both the source distribution $\psi(\zeta)$ and the kernel K (here representing the lensing potential) are unknown. This type of problem is generally unsolvable. In the case of lensing, however, one can use the knowledge that when the lensed image contains multiple distorted components of the background object these must arise from a common source. Furthermore, it is known that surface brightness is conserved. These 'priors' constrain the problem and permit the simultaneous solution of both the structure of the source and the properties of the lensing potential. This type of inversion problems for gravitational lenses has been developed extensively by Kochanek, Wallington and collaborators in several papers (cf. [80,79,148]). In particular, [79] developed an inversion method based on the CLEAN routine (cf. [68]) used in radio interferometry data reduction and applied it to PKS 1830-211. This method takes into consideration the effects of the finite resolution when attempting to invert the lens model and is thereby able to better distinguish the best lens model. The LensClean method of Kochanek & Narayan has produced the hitherto most reliable model for the PKS 1830-211 system, but due to the finite resolution, the inversion was done on radio data with rather low angular resolution but with good signal-to-noise, it is not likely to represent the final model.

Lehár et al. [87] modeled the PKS 1830-211 system using a singular isothermal elliptical mass distribution as well as with two singular isothermal spheres representing the lens galaxy and the source G2 (cf. Sect. 4.3.3). In both these cases they fixed the lens at the position of G, with a positional uncertainty of 80 mas. The extended radio emission was not used to constrain the lens model. Lehár et al. noted the strong dependence of the location of the lens galaxy and the Hubble constant derived from differential time delay measurements.

4.6.2   A New Lens Model of PKS 1830-211

The situation is rather unsatisfactory concerning the various solutions to the lensing configuration characterizing PKS 1830-211. Three different models ([79], [102,87]) give three different positions for the lens, with corresponding differences in the value of the Hubble constant for a given time delay. The situation is summarized in Fig. 4.25. The location of the cores in the long wavelength radio data used by Kochanek & Narayan [79] differ from that used by others (using data from shorter wavelengths). Their lens center relative to the other features shown in Fig. 4.25 is therefore somewhat uncertain.

The fact that surface brightness is conserved can be used in modeling a lensing configuration if both the observations and the code have infinitely good resolution. In the opposite extreme, if the source remains unresolved, one can

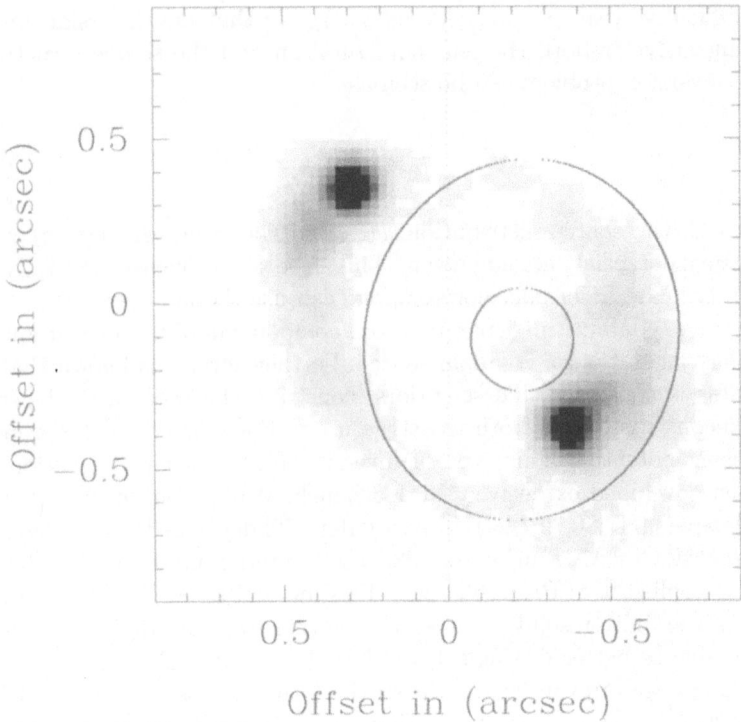

Fig. 4.26. A 15 GHz radio image of PKS 1830-211 (from [139]). The center of the coordinate system is arbitrary. All positions in the text and in Table 4.3 are relative the NE image. The critical lines of the best fit lens model, shown in Fig. 4.27 are shown for comparison.

calculate the magnification for a point source and apply a smoothing function (convolution) representing the observational transfer function (e.g. atmosphere, telescope, imaging array). A more difficult situation arises when an extended source distribution is partially resolved by the observer. A given resolution element will represent different areas of the source in a rather complicated manner and the conservation of surface brightness ceases to be a good prior for an observed lens system. In their LensClean method Kochanek & Narayan [79] solves this situation by representing the source emissivity distribution by δ-functions, mapped through the lens configuration with the corresponding magnification 'turned on' and then smoothed by a restoring beam.

An alternative way to solve the inversion problem as stated above is presented here. A more thorough description of the method will be published in Wiklind (2002). The concept is similar to the LensClean method of Kochanek & Narayan [79]. However, instead of introducing CLEAN components, as δ-functions to represent the source distribution, this method starts with a none-zero smooth source distribution and applies the Lucy rectification method [95] to constrain the source emissivity, given the observed lensed images and for a given lens

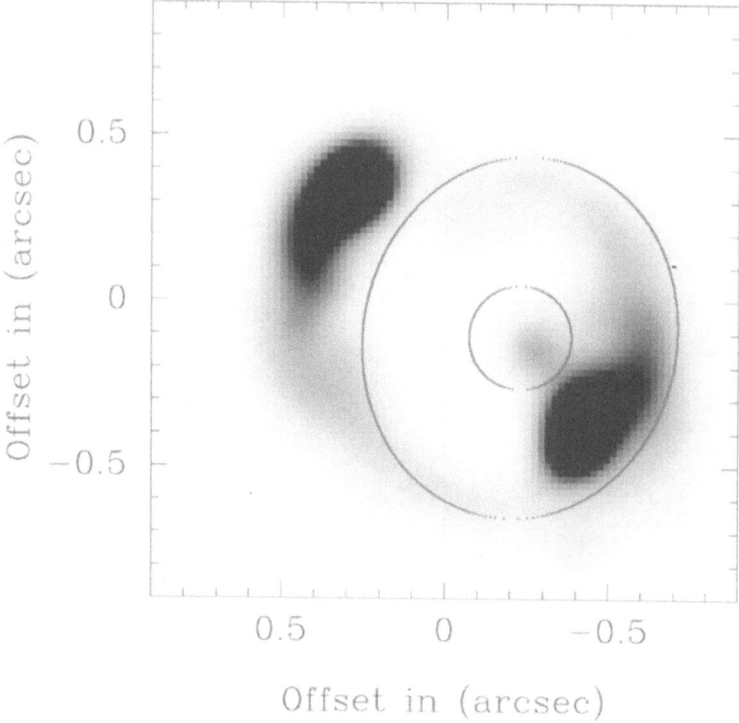

Offset in (arcsec)

Fig. 4.27. The lensed images obtained from inversion of the lens equation using Lucy rectification and Simulated Annealing (as described in the text). In this particular solution, the lens is fixed at the position of the observed (putative) lens center G (see Fig. 4.25). The critical lines from the lens model are marked.

model. The Lucy rectification scheme has been used extensively to deconvolve images obtained with the Hubble Space Telescope before the corrective optics was installed. It has also been used to deconvolve the molecular gas distribution in galaxies observed with single dish telescopes of rather poor angular resolution [156,158].

The method has an outer loop, which controls the lensing parameters, and an inner loop which solves for the best source distribution given the lens configuration. In the inner loop the source distribution is mapped through the lens and the resulting image is compared with the observed one. The source distribution is adjusted according to the Lucy method (see below) and the process is repeated. When the inner loop has converged, the source distribution is mapped through the lens a final time and the resulting image gives a goodness-of-fit. The lensing configuration is then modified in the outer loop and the process is repeated. The Lucy method is used only in the inner loop, each time starting with a perfectly smooth source distribution. The observed images are represented by

$$I_{obs}(x_1, x_2) = \int \int \psi(y_1, y_2) \, K(x_1, x_2 | y_1, y_2) \, dy_1 dy_2 \quad , \qquad (4.8)$$

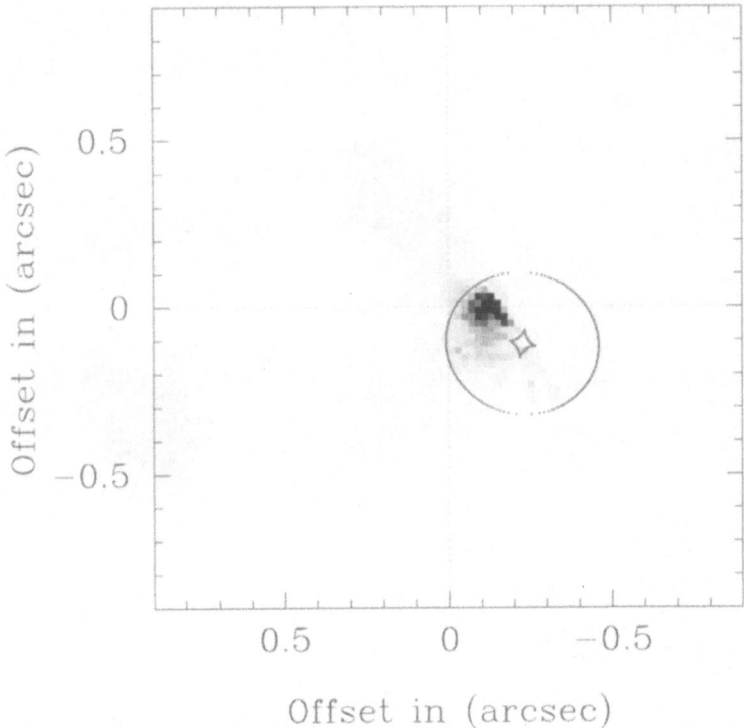

Offset in (arcsec)

Fig. 4.28. The source distribution derived from the best fit lens model, which gives the image seen in Fig. 4.27. The caustic structure is shown.

where $\psi(y_1, y_2)$ is the true source distribution and $K(x_1, x_2 | y_1, y_2)$ is the kernel representing both the gravitational lens and the finite angular resolution of the observation. The kernel is here written as a conditional probability function: the likelihood of (x_1, x_2) given (y_1, y_2). With this formulation we can use the Lucy method in a straightforward manner. The idea being that an approximation to the true source distribution is

$$\psi^{n+1}(y_1, y_2) = \psi^n(y_1, y_2) \int \int \frac{I_{obs}(x_1, x_2)}{I^n(x_1, x_2)} K(x_1, x_2 | y_1, y_2) \, dx_1 dx_2. \quad (4.9)$$

The conditional probability function K contains the likelihood of an image emissivity at position (x_1, x_2) given a source emissivity at (y_1, y_2) and the action of a restoring beam (i.e. a finite angular resolution). The simplest (although not entirely correct) restoring beam is a Gaussian with a HPBW similar to the angular resolution of the observations.

Even starting with a constant source emissivity distribution ψ^0, the Lucy rectification converges very rapidly to a specific source distribution. Unfortunately, there is no good criteria for determining when to stop the rectification (cf. [95]). In this particular application the inner loop was stopped after 5 iterations of the Lucy rectification. The outer loop consists only of changing the lensing parameters. Several different methods can be employed for this, the most efficient for this application being the Simplex method. However, there is a risk that this method gets stuck in a local minimum and great care has to be taken to ensure that a global minimum has really been reached. This involves repeatedly restarting the Simplex method with parameters offset from the ones giving a (local ?) minimum in the χ_ν^2. An alternative method that circumvents this, but that is computationally more expensive, is Simulated Annealing (cf. [99,156]). This latter method was used in this application.

The lens was modeled as a non-singular elliptical mass distribution

$$\kappa(x_1, x_2) = \kappa_0 \left(x_1^2 + \frac{x_2^2}{q^2} + s^2 \right)^{-\gamma} , \qquad (4.10)$$

where q is the projected axis ratio, s is the core radius and the surface density profile is set to $\gamma = 1/2$, representing an isothermal mass distribution. The deflection angle and magnification was calculated using the code developed by Barkana [8]. This code can handle surface density profiles with $\gamma \neq 1/2$, but here the modeling is restricted to the isothermal case. The density profile is, however, a very important parameter when deriving the Hubble constant using differential time delays (cf. [159,82]). In all there are six lens parameters that were fitted: the center position of the lens, the position angle, the velocity dispersion, the ellipticity, and the core radius.

Applying this method to the 15 GHz radio image of PKS 1830-211 shown in Fig. 4.26 [139], a best fit lens model is achieved with the parameters as tabulated in Table 4.3. Also listed in Table 4.3 are the results when keeping the lens position fixed at the coordinates of the putative infrared lens center G. The resulting image distribution for case (b) is seen in Fig. 4.27 together with the critical lines of the lens. The corresponding source distribution is seen in Fig. 4.28 together with the caustic structure.

The two solutions presented in Table 4.3 are very similar to each other, yet they give quite different values to the Hubble constant. Using the lens model of Nair et al. [102], the Hubble constant becomes $H_0 = 59^{+11}_{-8}$ km/s/Mpc. These differences are mainly due to the different locations of the lens center and introduces a large uncertainty in the correct value of H_0.

The lens model presented here will be further refined and the results should be regarded as tentative. However, unless the position of the lens can be derived more accurately, the value of the Hubble constant will remain uncertain. As mentioned above, the shape of the density profile is also a source of uncertainty for a more exact derivation of H_0. This uncertainty is largest for exponents $\gamma < 1/2$ [82].

4.7 Future Prospects

Existing millimeter and submillimeter telescopes use gravitational lenses more as an aid to the study of distant objects, rather than being an aid to the study of gravitational lensing as such. Nevertheless, some information about the content

Table 4.3. Lens parameters

	6 free parameters		4 free parameters (lens fixed)	
lens center[a]	-0.5008	-0.5205	-0.5010	-0.4450[b]
κ_0	0.3196		0.3106	
q	0.8576		0.8991	
s	0.0929		0.0745	
PA	135.553		101.174	
$\left(\frac{\Delta t}{28^{\text{days}}}\right)^{-1} H_0^c$	63^{+14}_{-6}		83^{+18}_{-9}	

(a) Relative to the NE component.

(b) Fixed at possible lens center G [87].

(c) With $\Delta t = 28^{+3}_{-5}$ days.

of the interstellar medium in both lenses and sources have been obtained, and the molecular absorption lines seen towards PKS 1830-211 have been used to measure the differential time delay in this particular system.

This situation will change dramatically when planned telescopes at millimeter and submillimeter wavelengths become available. Increased sensitivity and angular resolution will make this wavelength regime very important for studies of gravitational lenses as a phenomenon of their own. The most obvious advantage is that obscuration effects will be completely absent. The effects of microlensing will also be absent or at least minimal. The use of flux ratios of lensed components for constraining parameters when modeling lenses has fallen out of favor due to differential extinction and microlensing effects, but will be usable when the new submm/mm instruments are available.

Existing submillimeter and millimeter facilities include single dish telescopes, such as the IRAM 30m telescope on Pico Veleta in Spain, the JCMT 15m telescope on Mauna Kea, and the SEST 15m telescope on La Silla in Chile. Two 10m size dishes, aimed primarily for submillimeter wavelengths, include the CSO on Mauna Kea and the HHT on Mount Graham in Arizona. These telescopes use both heterodyne receivers for spectral line observations and bolometer type array cameras for continuum observations. The sensitivity depends largely on the quality of the site and the instrumentation. The angular resolution, however, is determined by the diffraction limit of the telescopes. At $\lambda = 1$mm, the angular resolution is limited to $10'' - 25''$. This constitutes the largest limitation to the study of gravitational lenses. For number counts (see Sect. 4.4.3) the lack of angular resolution means that with only slightly more sensitive receivers, confusion will become a major limitation (cf. [65])

A few interferometers operating at millimeter wavelengths exist. The IRAM Plateau de Bure interferometer in France consists of five (soon to be six) 15m telescopes, and represents the largest collecting area today. The OVRO interferometer consists of six 10m telescopes, while BIMA consists of eight 6m telescopes. In Japan the Nobeyama interferometer consists of six 10m telescopes. All of these facilities operate at $\lambda = 3 - 1\text{mm}$. The angular resolution reached is typically around $1''$ or slightly better. However, sensitivity becomes a serious limitation at the longest baselines and highest angular resolutions. Also, the Australian Telescope Compact Array (ATCA) has recently been upgraded to work at 3mm with five of its 22m elements.

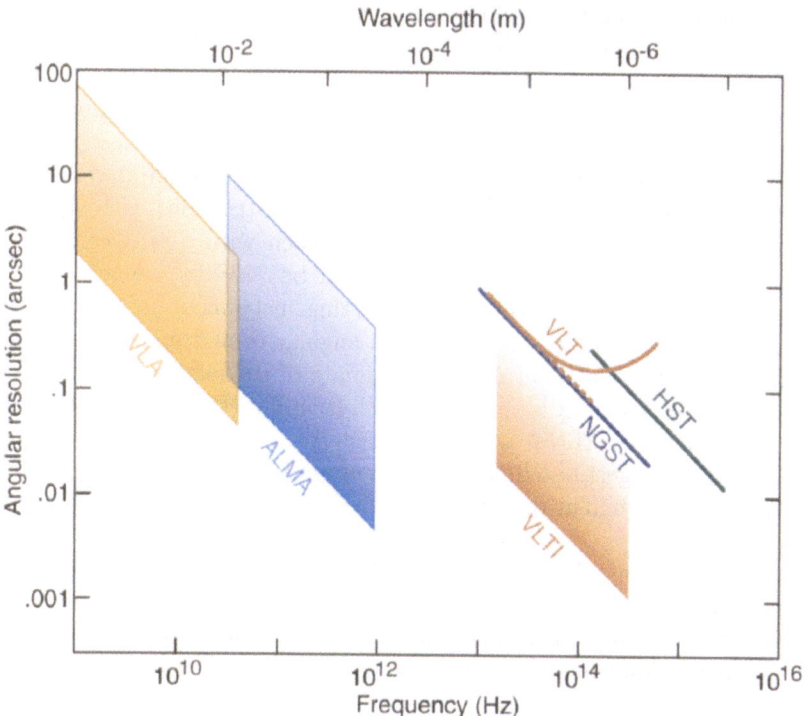

Fig. 4.29. A comparison of the wavelength coverage and the angular resolution of existing and planned instruments. The VLA refers to the Extended VLA. With ALMA, the same angular resolution will be reached from cm wavelengths into the optical (see Color Plate).

4.7.1   Future Instruments

Single Dish Telescopes. A few single dish submm/mm telescopes are under construction, or in advanced planning. These include the recently commissioned

Green Bank Telescope (GBT), which has a 90m unobstructed dish and will reach λ3mm when fully operational. The Large Millimeter Telescope is a 50m telescope built as a collaboration between INAOE and the University of Massachusetts in Amherst. This telescope will be placed on Sierra Negra in Mexico and operate at a wavelength of 1-4 mm. Two single dish telescopes to be situated close to the ALMA site (see below) are under construction: APEX is a 12m, single dish telescope which will be placed on Chajnantor at an altitude of 5000m. This telescope will operate into the THz regime, i.e. $\lambda \sim 300\mu m$. ASTE is a Japanese 10m dish, to be placed at Pampa La Bola, a few kilometers away from the ALMA site. The ASTE will also operate at submm to THz frequencies. These new single dish telescopes will explore distant objects, including gravitational lenses, with somewhat better angular resolution than existing telescopes. Nevertheless, even at the highest frequencies it will only reach an angular resolution of $\sim 6''$. This is insufficient for detailed studies of gravitationally lensed systems.

The Atacama Large Millimeter Array. A major step in submm/mm wave instruments will be the joint European-US project, with Japanese involvement as well, of building a large millimeter and submillimeter interferometer at an altitude of 5000m on Chajnantor in Chile. This instrument, with the acronym ALMA, will consist of $64 \times 12m$ telescopes, each with a surface accuracy of at least $20\mu m$. The total collecting area will be 7238 m^2, which is an order of magnitude greater than the largest existing instrument today. The longest baseline will be $10 - 12$ km, leading to an angular resolution surpassing that of the Hubble Space Telescope. A rough estimate of the angular resolution power is $0''2 \lambda_{mm}/L_{km}$, where λ is the wavelength in millimeters and L is the baseline length in kilometers. ALMA will easily reach $0''1$ resolution and, at the highest frequencies and longest baselines, $0''01$. A comparison of the wavelength coverage and the projected angular resolution for existing and planned telescopes is shown in Fig. 4.29.

At the same time ALMA will increase the sensitivity over existing instruments by at least two orders of magnitude. The noise rms level reached with an interferometer consisting of N array elements, each of effective area A_{eff} and with an integration time of t_{int} seconds over an effective bandwidth of B Hz, is expressed as

$$\Delta S = \frac{2k}{A_{eff}} \frac{T'_{sys} e^{\tau_\nu}}{\sqrt{N(N-1)\,B\,t_{int}}} \quad , \tag{4.11}$$

where the term e^τ represents the damping by the atmosphere at frequency ν. The system temperature, T'_{sys}, is the total noise received by the telescope (including ground pick-up). From this expression it is clear that large individual antenna sizes, a large number of array elements, a large bandwidth, a low system temperature and as little damping from the atmosphere are essential for a sensitive interferometer. These criteria can be fulfilled by using state-of-the-art receiver systems and putting them at a high and dry site such as Chajnantor in the Andes.

The excellent atmospheric conditions at Chajnantor are estimated to give a total system temperature (including the atmospheric damping) of ~230 K at a wavelength of 0.8mm. Using a 16 GHz wide backend and one hour of integration, the 5σ detection limit will be $100\,\mu Jy$. At a wavelength of $460\,\mu m$ the corresponding 5σ detection limit after one hour is 1 Jy. Although considerably worse, this wavelength range is usually not reachable at all with ground-based facilities. These values are representative for unresolved sources. If the source is resolved at the longest baselines, the sensitivity decreases.

Fig. 4.30. A simulated 9 arcmin^2 deep field observed with ALMA at a wavelength of $850\mu m$. All sources above the 5σ noise rms limit of 0.15 mJy are shown. The source population was modeled to fit the observed cumulative source counts, extrapolated to 0.1 mJy using a more shallow powerlaw than what can be fitted in the range 1-10 mJy. **Left:** Unlensed field. Approximately 210 sources are 'detected'. **Right:** The same field as in the left image but with an intervening rich galaxy cluster at redshift $z = 0.3$.

4.7.2   Weak Lensing at Submillimeter Wavelengths

With its superb sensitivity and angular resolution, ALMA can successfully be used to study weak lensing by intermediate redshift galaxy clusters. Assuming a constant co-moving volume density of dusty high redshift objects, the constant sensitivity to dust continuum emission from $z \approx 1$ to $z \approx 10$ (Sect. 4.4.2) favors the detection of the highest redshift objects. This selection bias is unique for the submm/mm wavelength regime. Furthermore, a reasonably well-sampled uv-plane, resulting from a large number of interferometer elements, will have a point-spread function (PSF) which is well behaved in comparison to existing wide-field CCDs. And, finally, the dust distribution in galaxies tend to be symmetric and centrally concentrated, leading to a simple geometry of the lensed sources. The latter issue is, of course, not well constrained by existing data. Although dust continuum emission will be easier to detect at large distances than line emission (cf. [32]), the high sensitivity and broad instantaneous bandwidth of ALMA will allow CO lines to be observed for many of the lensed sources. This enables a direct way of deriving the redshift distribution of the source population. Furthermore, observations of CO lines in the intervening galaxies will allow a determination of the dynamical mass of the lenses.

ALMA will, however, have a relatively limited instantaneous field of view. This is particularly aggravating at high frequencies. The field of view is limited by the size of the individual telescopes and will be similar in size to that of existing single dish telescopes. This means that the time consuming process of mosaicing is necessary. The key parameters for the effectiveness or speed of this new instrument are the sensitivity ΔS, the angular resolution and the primary beam area A_{fov}. The time required to survey a given area A to a flux density limit S is then: $t \approx (\Delta S/S)^2 (A/A_{fov})$ (cf. [20] [17]).

An example of what can be achieved with ALMA is shown in Fig. 4.30. The left panel shows a simulated 9 arcmin^2 field observed with ALMA to a 5σ noise rms limit of 0.15 mJy. The wavelength is 0.8mm and was chosen to maximize the number of detected dust continuum sources while minimizing the observing time (by making the field-of-view as large as possible). The field of view per pointing is $\sim$0.07 arcmin^2. As shown in Sect. 4.7.1 it takes somewhat less than one hour of telescope time to reach this 5σ detection limit. To cover 9 arcmin^2 thus requires a total of 130 hours. In this example more than 200 sources will be detected (left panel) and their distortion by a rich cluster at $z = 0.3$ will be clearly detectable (right panel). In this example a population of dusty galaxies was assumed to have a constant co-moving volume density between $z = 1 - 7$ and undergo pure luminosity evolution. The observed flux densities were fitted to the number counts derived from SCUBA and MAMBO observations (Sect. 4.4.3). The median luminosity of the 'detected' galaxies is 3×10^{11} L$_\odot$ and their redshift distribution is more or less flat between $z = 2$ to $z = 7$. The cumulative source counts was not, however, extrapolated to weaker flux densities using the same powerlaw as applicable between 10-1 mJy (cf. Fig. 4.8, but with a more modest slope. This ensures that the number of sources seen in the right panel of Fig. 4.30 is not an overestimate.

In the same manner as lensing by intermediate redshift clusters is presently being used to observe dust continuum sources at flux density levels otherwise not reachable with existing instruments, ALMA will reach flux densities approaching μJy levels when observing lensed sources. For high redshift galaxies, this flux density level corresponds to luminosities around 10^9 L$_\odot$, i.e. dwarf galaxies.

4.8   Summary

The study of gas and dust at high redshift is important for several reasons. It gives us an unbiased view of star formation activity in obscured objects and it tells the story of the chemical evolution and star formation history in galaxies through the amount of processed gas (and dust) it contains. With today's millimeter and submillimeter facilities, this research area has used gravitational lensing mostly as a tool to boost the sensitivity. This is evident through the preponderance of gravitationally lensed objects among those which have been detected at $z > 2$ in the lines of the CO molecule. It is also evident in the use of lensing magnification by galaxy clusters in order to reach faint submm/mm continuum sources. There are, however, a few cases where millimeter lines have been directly involved in understanding lensing configurations. The best example of this is the highly obscured PKS 1830-211, where the lens was identified through molecular absorption lines and where these lines give a velocity dispersion measure by originating in two different regions of the lens. The molecular absorption lines in this system have also been used to derive the differential time delay between the two main components, the main objective being to determine the Hubble constant, but also adding to the constraints in modeling this particular lens system.

With future millimeter and submillimeter instruments, such as ALMA, coming on-line, the situation is likely to change drastically. The sensitivity of ALMA will be such that it does not need the extra magnification from lensing to observe very distant objects. Instead it will be used to study the lensing itself. The more or less constant sensitivity to dust emission over a redshift range stretching from $z \approx 1$ to $z \approx 10$ means that the likelihood for strong lensing of dust continuum detected sources is much larger than for optically selected sources. ALMA will therefore discover many more lenses and allow a direct assessment of cosmological parameters through lens statistics. Weak lensing will also be an area where ALMA can successfully contribute. Again, the high sensitivity to dust emission out to very high redshifts, combined with an angular resolution $< 0.''1$, and a more beneficial 'PSF' will make ALMA more efficient for probing the potential of galaxy clusters than present day optical/IR telescopes. In addition we will be able to study both the sources and the lenses themselves, free of obscuration and extinction corrections, derive rotation curves for the lenses, their orientation and, thus, greatly constrain lens models.

Acknowledgments

T.W. thanks Françoise Combes for allowing the use of unpublished material on millimeter wave absorption line systems. Many thanks to F. Combes, D. de Mello, P. Cox and F. Courbin for careful reading of the manuscript and for valuable comments.

References

1. D. Alloin, R. Barvainis, M.A. Gordon, R. Antonucci: A&A **265**, 429 (1992)
2. D. Alloin, R. Barvainis, S. Guilloteau: ApJ **528**, L81 (2000)
3. D. Alloin, S. Guilloteau, R. Barvainis, R. Antonucci, L. Tacconi: A&A **321**, 24 (1997)
4. P. Andreani, A. Cimatti, L. Loinard, H. Röttgering: A&A **354**, L1 (2000)

5. A.J. Barger, L.L. Cowie, D.B. Sanders, E. Fulton, Y. Taniguchi, Y. Sato, K. Kawara, H. Okuda: Nature **394**, 248 (1998)
6. A.J. Barger, L.L. Cowie, I. Smail, R.J. Ivison, A.W. Blain, J.-P. Kneib: AJ **117**, 2656 (1999)
7. A.J. Barger, L.L. Cowie, E.A. Richards: AJ **119**, 2029 (2000)
8. R. Barkana: ApJ **502**, 531 (1998)
9. R. Barvainis, D. Alloin, R. Antonucci: ApJ **337**, L69 (1989)
10. R. Barvainis, D. Alloin, M. Bremer: A&A in press, (2002)
11. R. Barvainis, D. Alloin, S. Guilloteau, R. Antonucci R.: ApJ **492**, L13 (1998)
12. R. Barvainis, L. Tacconi, R. Antonucci, D. Alloin, P. Coleman: Nature **371**, 586 (1994)
13. R. Barvainis, P. Maloney, R. Antonucci, D. Alloin: ApJ **484**, 695 (1997)
14. E. Bertin, S. Arnouts: A&AS **117**, 393 (1996)
15. P.R. Bevington, D.K. Robinson: *Data reduction and error analysis for the physical sciences*, New York, McGraw-Hill, 2nd edition, (1992)
16. A.W. Blain: MNRAS **304**, 669 (1999)
17. A.W. Blain, M.S. Longair: MNRAS **279**, 847 (1996)
18. A.W. Blain, J.-P. Kneib, R.J. Ivison, I. Smail: ApJ **512**, L87 (1999a)
19. A.W. Blain, I. Smail, R.J. Ivison, J.-P. Kneib: MNRAS **302**, 632 (1999b)
20. A.W. Blain: in *Science with the Atacama Large Millimeter Array (ALMA)*, (Associated Universities Inc., Washington D.C., 1999d)
21. C. Borys, S.C. Chapman, M. Halpern, D. Scott: MNRAS, **330**, L63 (2002)
22. T.J. Broadhurst, J. Lehár: ApJ **450**, L41 (1995)
23. R.L. Brown, P.A. Vanden Bout: ApJ **397**, L19 (1992)
24. I.A.W. Browne, A.R. Patnaik, D. Walsh, P.N. Wilkinson: MNRAS **263**, L32 (1993)
25. C.L. Carilli, E.S. Perlman, J.T. Stocke: ApJ **400**, L13 (1992)
26. C.L. Carilli, M.P. Rupen, B. Yanny: ApJ **412**, L59 (1993)
27. C.L. Carilli, K.M. Menten, M.J. Reid, M.P. Rupen: ApJ **474**, L89 (1997a)
28. C.L. Carilli, K.M. Menten, M.J. Reid, M.P. Rupen, M. Claussen: 13th IAP Colloqium: *Structure and Evolution of the IGM from QSO Absorption Line Systems*, eds. P. Petitjean, S. Charlot (1997b)
29. C.L. Carilli, M.S. Yun: ApJ **530**, 618 (2000)
30. C.L. Carilli, F. Bertoldi, A. Bertarini, K.M. Menten, E. Kreysa, R. Zylka, F. Owen, M. Yun: (2000), astro-ph/0009298
31. F. Combes: Ap&SS **269**, 405 (1999)
32. F. Combes, R. Maoli, A. Omont: A&A **345**, 369 (1999)
33. F. Combes, T. Wiklind: A&A **303**, L61 (1995)
34. F. Combes, T. Wiklind: in *Cold Gas at High Redshift*, Bremer, M., van der Werf, P., Carilli, C. (eds.), Kluwer, (1996)
35. F. Combes, T. Wiklind: ApJ **486**, L79 (1997)
36. F. Courbin, C. Lidman, B.L. Frye, P. Magain, T.J. Broadhurst, M.A. Pahre, S.G. Djorgovski: ApJ **499**, L119 (1998)
37. F. Courbin, G. Meylan, J.-P. Kneib, C. Lidman: ApJ, in press (2002)
38. P. Cox, A. Omont, S.G. Djorgovski et al.: A&A **387**, 406 (2002)
39. S.G. Djorgovski, et al.: MNRAS **257**, 240 (1992)
40. D. Downes, P.M. Solomon, S.J.E. Radford: ApJ **453**, L65 (1995)
41. D. Downes, P.M. Solomon: ApJ **507**, 615 (1998)
42. D. Downes D., R. Neri, A. Greve, S. Guilloteau, F. Casoli, D. Hughes, etal.: A&A **347**, 809 (1999a)

43. D. Downes, R. Neri, T. Wiklind, D.J. Wilner, P.A. Shaver: ApJ **513**, L1 (1999b)
44. S. Eales, S. Lilly, W. Gear, L. Dunne, J.R. Bond, F. Hammer, O. Le Fèvre, D. Crampton: ApJ **515**, 518 (1999)
45. S. Eales, S. Lilly, T. Webb, L. Dunne, W. Gear, D. Clements, M. Yun: AJ **120**, 2244 (2000)
46. R.A. Edelson, J.H. Krolik: ApJ **333**, 646 (1988)
47. E. Egami, G. Neugebauer, B.T. Soifer, K. Matthews, M. Ressler, E.E. Becklin, T.W. Murphy Jr., D.A. Dale: ApJ **535**, 561 (2000)
48. S.L. Ellison, G.F. Lewis, M. Pettini, F.H. Chaffee, M.J. Irwin: ApJ **520**, 456 (1999)
49. A.S. Evans: 1999, in ASP Conf. Series **156** (1999)
50. A.C. Fabian, I. Smail, K. Iwasawa, S.W. Allen, A.W. Blain, C.S. Crawford, et al.: MNRAS **315**, L8 (2000)
51. C.D. Fassnacht, T.J. Pearson, A.C.S. Readhead, I.W.A. Browne, L.V.E. Koopmans, S.T. Myers, P.N. Wilkinson: ApJ **527**, 498 (1999)
52. D.J. Fixsen, E. Dwek, J.C. Mather, C.L. Bennett, R.A. Shafer: ApJ **508**, 123 (1998)
53. D.T. Frayer, R.J. Ivison, N.Z. Scoville, M.S. Yun, A.S. Evans, I. Smail, A.W. Blain, J.-P. Kneib: ApJ **506**, L7 (1998)
54. D.T. Frayer, R.J. Ivison, N.Z. Scoville, A.S. Evans, M.S. Yun, I. Smail, A.J. Barger, A.W. Blain, J.-P. Kneib: ApJ **514**, L13 (1999)
55. D.T. Frayer, I. Smail, R.J. Ivison, N.Z. Scoville: AJ **120**, 1668 (2000)
56. M. Gerin, T.G. Phillips, D.J. Benford, K.H. Young, K.M. Menten, B.L. Frye: ApJ **488**, 31 (1997)
57. R. Gispert, G. Lagache, J.-L. Puget: A&A **360**, 1 (2000)
58. G.L. Granato, L. Danese, A. Franceschini: ApJ **460**, L11 (1996)
59. G.L. Granato, L. Danese, A. Franceschini: ApJ **486**, 147 (1997)
60. S. Guilloteau, A. Omont, R. McMahon, P. Cox, P. Petitjean: A&A **328**, L1 (1997)
61. S. Guilloteau, A. Omont, P. Cox, R.G. McMahon, P. Petitjean: A&A **349**, 363 (1999)
62. D.B. Haarsma, R.B. Partridge, R.A. Windhorst, E.A. Richards: ApJ **544**, 641 (2000)
63. J.N. Hewitt, E.L. Turner, C.R. Lawrence, D.P. Schneider, J.P. Brody: AJ **104**, 968 (1992)
64. R.H. Hildebrand: QJRAS **24**, 267 (1983)
65. D.W. Hogg: AJ **121**, 1207 (2001)
66. D.H. Hughes, J.S. Dunlop, S. Rawlins: MNRAS **289**, 766 (1997)
67. D.H. Hughes, et al.: Nature **394**, 241 (1998)
68. J.A. Högbom: A&AS **15**, 417 (1974)
69. R.A. Ibata, G.F. Lewis, M.J. Irwin, J. Lehár, E.J. Totten: AJ **118**, 1922 (1999)
70. M.J. Irwin, R.A. Ibata, G.F. Lewis, E.J. Totten: ApJ **505**, 529 (1998)
71. R.J. Ivison, I. Smail, J.-F. Le Borgne, A.W. Blain, J.-P. Kneib, et al.: MNRAS **298**, 583 (1998)
72. R.J. Ivison, I. Smail, A.J. Barger, J.-P. Kneib, A.W. Blain, F.N. Owen, T.H. Kerr, L.L. Cowie: MNRAS **315**, 209 (2000)
73. R.J. Ivison, I. Smail, D.T. Frayer, J.-P. Kneib, A.W. Blain: ApJ **561**, L45 (2001)
74. D.L. Jauncey: ARAA **13**, 23 (1975)
75. D.L. Jauncey, J.E. Reynolds, A.K. Tzioumis, T.W.B. Muxlow, R.A. Perley, D.W. Murphy, et al.: Nature **352**, 132 (1991)
76. R. Kayser, J. Surdej, J.J. Condon, K.I. Kellermann K., P. Magain, M. Remy, A. Smette: ApJ **364**, 15 (1990)

77. J.-P. Kneib, D. Alloin, Y. Mellier, S. Guilloteau, R. Barvainis, et al.: A&A **329**, 827 (1998a)

78. J.-P. Kneib, D. Alloin, R. Pello: A&A **339**, L65 (1998b)

79. C.S. Kochanek, R. Narayan: ApJ **401**, 461 (1992)

80. C.S. Kochanek, R.D. Blandford, C.R. Lawrence, R. Narayan R.: MNRAS **238**, 43 (1989)

81. L.V.E. Koopmans, A.G. de Bruyn: A&A **358**, 793 (2000)

82. L.V.E. Koopmans, C.D. Fassnacht: ApJ **527**, 513 (1999)

83. G. Lagache, J.-L. Puget, R. Gispert: Ap&SS **269**, 263 (1999)

84. T.R. Lauer: ApJ **292**, L104 (1985)

85. C. Ledoux, B. Théodore, P. Petitjean, M.N. Bremer, G.F. Lewis, R.A. Ibata, M.J. Irwin, E.J. Totten: A&A **339**, L77 (1998)

86. J. Lehár, B. Burke, S. Conner, E.E. Falco, A. Fletcher, et al.: AJ **114**, 48 (1997)

87. J. Lehár, E.E. Falco, C.S. Kochanek, B.A. McLeod, J.A. Muños, C.D. Impey, H.-W. Rix, C.R. Keeton, C.Y. Peng C. Y.: ApJ **536**, 584 (2000)

88. M. Lerner M., L. Bååth L., M. Inoue M., et al.: A&A **280**, 117 (1993)

89. G.F. Lewis, S.C. Chapman, R.A. Ibata, M.J. Irwin, E.J. Totten: ApJ **505**, L1 (1998)

90. G.F. Lewis, M. Russell, R.A. Ibata: PASP **111**, 1503 (1999)

91. G.F. Lewis, C. Carilli, P. Papadopoulos, R.J. Ivison: MNRAS, **330**, L15 (2002)

92. C. Lidman, F. Courbin, G. Meylan, T. Broadhurst, B.L. Frye, W. Welch: ApJ **514**, L57 (1998)

93. J.E.J. Lovell, J.E. Reynolds, D.L. Jauncey, et al.: ApJ **472**, L5 (1996)

94. J.E.J. Lovell, D.L. Jauncey, J.E. Reynolds, M.H. Wieringa, E.A. King, A.K. Tzioumis, P.M. McCulloch, P.G. Edwards: ApJ **508**, L51 (1998)

95. L.B. Lucy: AJ **79**, 745 (1974)

96. P. Magain, J. Surdej, J.-P. Swings, U. Borgeest, R. Kayser, et al.: Nature **334**, 325 (1988)

97. K.M. Menten, C.L. Carilli, M.J. Reid: in *Highly Redshifted Radio Lines*, C. L. Carilli, K. M. Menten, G. Langston (eds.), ASP: San Francisco (1999)

98. K.M. Menten, M.J. Reid: ApJ **465**, L99 (1996)

99. N. Metropolis, A.W. Rosenbluth, M.N. Rosenbluth, A.H. Teller, E. Teller: J. Chem. Phys. **21**, 1087 (1953)

100. J.M. Mazzarella, J.R. Graham , D.B. Sanders, S. Djorgovski: ApJ **409**, 170 (1993)

101. I.F. Mirabel, D.B. Sanders, I. Kazes: ApJ **340**, L9 (1989)

102. S. Nair, D. Narashima, A.P. Rao: ApJ **407**, 46 (1993)

103. K. Ohta, T. Yamada, K. Nakanishi, K. Khono, M. Akiyama, et al.: Nature **382**, 426 (1996)

104. A. Omont, P. Cox, F. Bertoldi, R.G. McMahon, C. Carilli, K.G. Isaak: A&A **374**, 371 (2001)

105. A. Omont, P. Petitjean, S. Guilloteau, R. McMahon, P.M. Solomon, et al.: Nature **382**, 428 (1996)

106. P. Papadopoulos, R. Ivison, C. Carilli, G. Lewis: Nature **409**, 58 (2001)

107. P.P. Papadopoulos, H.J.A., Röttgering, P.P. van der Werf, S. Guilloteau, A. Omont, W.J.M. van Breugel: ApJ **528**, 626 (2000)

108. A.R. Patnaik, I.W.A. Browne, L.J. King, T.W.B. Muxlow, D. Walsh, P.N. Wilkinson: MNRAS **261**, 435 (1993)

109. A.R. Patnaik, R.W. Porcas, I.W.A. Browne: MNRAS **274**, L5 (1995)

110. J. Pelt, W. Hoff, R. Kayser, S. Refsdal, T. Schramm: A&A **256**, 775 (1994)

111. J. Pelt, R. Kayser, S. Refsdal, T. Schramm: A&A **305**, 97 (1996)

112. J. Pelt, J. Hjorth, S. Refsdal, R. Schild, R. Stabell: A&A **337**, 681 (1998)
113. E.S. Perlman, C.L. Carilli, J.T. Stocke, J. Conway: AJ **111**, 1839 (1996)
114. T.G. Phillips, B.N. Ellison, J.B. Keene, R.B. Leighton, R.J. Howard, C.R. Masson, et al.: ApJ **322**, L73 (1987)
115. M. Polletta, T.J.-L. Courvoisier: A&A **350**, 765 (1999)
116. K.M. Pontoppidan: Strongly Lensed high-z ULIRGs?, Master Thesis at the University of Copenhagen (2000)
117. W.H. Press, G.B. Rybicki, J.N. Hewitt: ApJ **385**, 416 (1992)
118. J.-L. Puget, A. Abergel, J.-P. Bernard, F. Boulanger, W.B. Burton, F.-X. Desert, D. Hartmann: A&A **308**, L5 (1996)
119. S.J.E. Radford, P.M. Solomon, D. Downes: ApJ **369**, L15 (1991)
120. A.P. Rao, R. Subrahmanyan: MNRAS **231**, 229 (1988)
121. M. Rowan-Robinson, A. Efstathiou: MNRAS **263**, 675 (1993)
122. M. Ryle: ARAA **6**, 249 (1967)
123. D.B. Sanders, J.S. Young, N.Z. Scoville, B.T. Soifer, G.E. Danielson G.: ApJ **312**, L5 (1987)
124. D.B. Sanders, N.Z. Scoville, B.T. Soifer: 1988, ApJ **335**, L1 (1988)
125. D.B. Sanders, N.Z. Scoville, A. Zensus, B.T. Soifer, T.L. Wilson, R. Zylka, H. Steppe: A&A **213**, L5 (1989)
126. P. Schneider, J. Ehlers, E.E. Falco: *Gravitational lenses*, Springer Verlag, New York (1992)
127. N.Z. Scoville, S. Padin, D.B. Sanders, B.T. Soifer, M.S. Yun: ApJ **415**, L75 (1993)
128. N.Z. Scoville, M.S. Yun, R.L. Brown, P.A. Vanden Bout: ApJ **449**, L101 (1995)
129. N.Z. Scoville, M.S. Yun, R.A. Windhorst, W.C. Keel, L. Armus: ApJ **485**, L21 (1997)
130. R. Siebenmorgen, E. Krügel, V. Zota: A&A **351**, 140 (1999)
131. S.E. Scott, M.J. Fox, J.S. Dunlop, et al.: MNRAS **331**, 817 (2002)
132. I. Smail, R.J. Ivison, A.W. Blain: ApJ **490**, L5 (1997)
133. I. Smail, A.C. Edge, R.C. Ellis, R.D. Blandford: MNRAS **293**, 124 (1998a)
134. I. Smail, R.J. Ivison, J.-P. Kneib: ApJ **507**, L21 (1998b)
135. I. Smail, R.J. Ivison, F.N. Owen, A.W. Blain, J.-P. Kneib: ApJ **528**, 612 (2000)
136. P.M. Solomon, D. Downes, S.J.E. Radford: ApJ **398**, L29 (1992)
137. P.M. Solomon, D. Downes, S.J.E. Radford, J.W Barrett: ApJ **478**, 144 (1997)
138. P.M. Solomon, A.R. Rivolo, J. Barrett, A. Yahil: ApJ **319**, 730 (1987)
139. R. Subrahmanyan, D. Narashima, A.P. Rao, G. Swarup: MNRAS **246**, 263 (1990)
140. J.J. Swift, W.J. Welch, B.L. Frye: ApJ **549**, L29 (2001)
141. T.T. Takeuchi, R. Kawabe, K. Kohno, K. Nakanishi, T.T. Ishii, H. Hirashita, K. Yoshikawa: PASJ **53**, 381 (2001)
142. H.A. Thronson, C.M. Telesco: ApJ **311**, 98 (1986)
143. G. Thuma, N. Neininger, U. Klein, R. Wielebinski: A&A **358**, 65 (2000)
144. J. Tonry, C.S. Kochanek: AJ **117**, 2034 (1999)
145. M. Tsuboi, N. Nakai: PASJ **45**, L179 (1994)
146. D. Turnshek, O. Lupie, S. Rao, B. Espey, C. Sirola: ApJ **476**, 40 (1997)
147. T. van Ommen, D. Jones, R. Preston, D. Jauncey: ApJ **444**, 561 (1995)
148. S. Wallington, C.S. Kochanek, R. Narayan: ApJ **465**, 64 (1996)
149. T. Wiklind, F. Combes: A&A **286**, L9 (1994)
150. T. Wiklind, F. Combes: A&A **299**, 382 (1995)
151. T. Wiklind, F. Combes: Nature **379**, 139 (1996a)
152. T. Wiklind, F. Combes: A&A **315**, 86 (1996b)
153. T. Wiklind, F. Combes: A&A **328**, 48 (1997)

154. T. Wiklind, F. Combes: ApJ **500**, 129 (1998)
155. T. Wiklind, F. Combes: in *Highly Redshifted Radio Lines*, ASP Conf. Series Vol. 156, eds. C. L. Carilli, K. M. Menten, G.I. Langston, p. 202 (1999)
156. T. Wiklind, C. Henkel:A&A **257**, 437 (1992)
157. T. Wiklind, C. Henkel: A&A **297**, L71 (1995)
158. T. Wiklind, C. Henkel, L.J. Sage: A&A **271**, 71 (1993)
159. L.L.R. Williams, P. Saha: ApJ **119**, 439 (2000)
160. D.J. Wilner, J.-H. Zhao, P.T.P. Ho: ApJ **453**, L91 (1995)
161. R.A. Windhorst, G.K. Miley, F.N. Owen, R.G. Kron, D.C. Koo: ApJ **289**, 494 (1985)
162. W. Xu, A.C.S. Readhead, T.J. Pearson, A.G. Polatidis, P.N. Wilkinson: ApJS **99**, 297 (1995)
163. T. Yamada, K. Ohta, A. Tomita, T. Takata: AJ **110**, 1564 (1995)
164. M.S. Yun, N.Z. Scoville, J.J. Carrasco, R.D. Blandford: ApJ **479**, L9 (1997)

Subject Index

absorption lines
- at high redshift, 133
- by lensing galaxies, 135
- channel maps, 138
- in lensing galaxies, 137
- in quasar host galaxies, 136
- integrated opacity, 134
- macroscopic absorption coefficient, 142
- molecular, 133, 161
ACS, 46, 121
ALMA, 127, 151, 180–183
anisotropy parameter, 108, 111, 112
arclets, 25, 46, 70
arcs, 25, 45, 46, 70
arrival time, 1–7, 12, 16, 17, 45

BIMA, 127
bolometer arrays, 145

CASTLES, 39, 47
caustics, 6–9, 14, 58, 150, 160
caustics network, 16, 30
cloverleaf, 157, 158
cluster
- as a lens, 69, 71, 72, 145, 183
- bias, 77
- bias factor, 115
- dark, 89, 90
- dynamics, 70
- high redshift, 79
- mass function, 78, 79
- mass sheet, 33, 34
- mass-to-light ratio, 76–78, 80
- survey, 81
- velocity dispersion, 80
- X-ray, 26, 63, 70, 72, 74, 82
CMB, 74, 91
- temperature, 133
compactness, 4

convergence, 4, 6, 13, 33, 57, 59, 74, 81, 98, 107
conversion factor, 128
core, 8, 17
- radius, 17
cosmology, 16, 19, 45, 59, 84, 85, 90, 97
covering factor, 13
critical curves, 6–9, 14, 17, 176
curvature, 5, 6, 12

dark energy, 90
dark halos
- as sources, 116
- extent, 102, 104, 114
- field galaxies, 102
- flattening, 97, 106, 109
- model, 103
- shape, 97, 106
- truncation, 106, 111
- velocity dispersion, 103, 104
dark matter, 55, 77, 79, 89, 96
degeneracy, 16, 40
- central concentration, 37, 38
- mass disk, 17, 18, 37
- mass sheet, 58, 59, 71, 74
demagnification, 6, 7, 12
density
- critical, 4, 13, 14, 16, 57, 76
- profile, 32, 97, 177
- surface, 4, 13, 40, 57, 98
depletion curve, 60, 61
distances, 23
- angular diameter, 2, 59, 141
dust
- blackbody emission, 141
- cold, 124, 153
- continuum emission, 140, 143, 154
- extinction, 163, 164
- high redshift, 181, 182

Druck: betz-druck GmbH, D-64291 Darmstadt
Verarbeitung: Buchbinderei Schäffer, D-67269 Grünstadt

Lecture Notes in Physics

For information about Vols. 1–562
please contact your bookseller or Springer-Verlag

Monographs

For information about Vols. 1–29
please contact your bookseller or Springer-Verlag

Vol. m 30: A. J. Greer, W. J. Kossler, Low Magnetic Fields in Anisotropic Superconductors. VII, 161 pages. 1995.

Vol. m 31 (Corr. Second Printing): P. Busch, M. Grabowski, P.J. Lahti, Operational Quantum Physics. XII, 230 pages. 1997.

Vol. m 32: L. de Broglie, Diverses questions de mécanique et de thermodynamique classiques et relativistes. XII, 198 pages. 1995.

Vol. m 33: R. Alkofer, H. Reinhardt, Chiral Quark Dynamics. VIII, 115 pages. 1995.

Vol. m 34: R. Jost, Das Märchen vom Elfenbeinernen Turm. VIII, 286 pages. 1995.

Vol. m 35: E. Elizalde, Ten Physical Applications of Spectral Zeta Functions. XIV, 224 pages. 1995.

Vol. m 36: G. Dunne, Self-Dual Chern-Simons Theories. X, 217 pages. 1995.

Vol. m 37: S. Childress, A.D. Gilbert, Stretch, Twist, Fold: The Fast Dynamo. XI, 406 pages. 1995.

Vol. m 38: J. González, M. A. Martín-Delgado, G. Sierra, A. H. Vozmediano, Quantum Electron Liquids and High-Tc Superconductivity. X, 299 pages. 1995.

Vol. m 39: L. Pittner, Algebraic Foundations of Non-Com-mutative Differential Geometry and Quantum Groups. XII, 469 pages. 1996.

Vol. m 40: H.-J. Borchers, Translation Group and Particle Representations in Quantum Field Theory. VII, 131 pages. 1996.

Vol. m 41: B. K. Chakrabarti, A. Dutta, P. Sen, Quantum Ising Phases and Transitions in Transverse Ising Models. X, 204 pages. 1996.

Vol. m 42: P. Bouwknegt, J. McCarthy, K. Pilch, The W3 Algebra. Modules, Semi-infinite Cohomology and BV Algebras. XI, 204 pages. 1996.

Vol. m 43: M. Schottenloher, A Mathematical Introduction to Conformal Field Theory. VIII, 142 pages. 1997.

Vol. m 44: A. Bach, Indistinguishable Classical Particles. VIII, 157 pages. 1997.

Vol. m 45: M. Ferrari, V. T. Granik, A. Imam, J. C. Nadeau (Eds.), Advances in Doublet Mechanics. XVI, 214 pages. 1997.

Vol. m 46: M. Camenzind, Les noyaux actifs de galaxies. XVIII, 218 pages. 1997.

Vol. m 47: L. M. Zubov, Nonlinear Theory of Dislocations and Disclinations in Elastic Body. VI, 205 pages. 1997.

Vol. m 48: P. Kopietz, Bosonization of Interacting Fermions in Arbitrary Dimensions. XII, 259 pages. 1997.

Vol. m 49: M. Zak, J. B. Zbilut, R. E. Meyers, From Instability to Intelligence. Complexity and Predictability in Nonlinear Dynamics. XIV, 552 pages. 1997.

Vol. m 50: J. Ambjørn, M. Carfora, A. Marzuoli, The Geometry of Dynamical Triangulations. VI, 197 pages. 1997.

Vol. m 51: G. Landi, An Introduction to Noncommutative Spaces and Their Geometries. XI, 200 pages. 1997.

Vol. m 52: M. Hénon, Generating Families in the Restricted Three-Body Problem. XI, 278 pages. 1997.

Vol. m 53: M. Gad-el-Hak, A. Pollard, J.-P. Bonnet (Eds.), Flow Control. Fundamentals and Practices. XII, 527 pages. 1998.

Vol. m 54: Y. Suzuki, K. Varga, Stochastic Variational Approach to Quantum-Mechanical Few-Body Problems. XIV, 324 pages. 1998.

Vol. m 55: F. Busse, S. C. Müller, Evolution of Spontaneous Structures in Dissipative Continuous Systems. X, 559 pages. 1998.

Vol. m 56: R. Haussmann, Self-consistent Quantum Field Theory and Bosonization for Strongly Correlated Electron Systems. VIII, 173 pages. 1999.

Vol. m 57: G. Cicogna, G. Gaeta, Symmetry and Perturbation Theory in Nonlinear Dynamics. XI, 208 pages. 1999.

Vol. m 58: J. Daillant, A. Gibaud (Eds.), X-Ray and Neutron Reflectivity: Principles and Applications. XVIII, 331 pages. 1999.

Vol. m 59: M. Kriele, Spacetime. Foundations of General Relativity and Differential Geometry. XV, 432 pages. 1999.

Vol. m 60: J. T. Londergan, J. P. Carini, D. P. Murdock, Binding and Scattering in Two-Dimensional Systems. Applications to Quantum Wires, Waveguides and Photonic Crystals. X, 222 pages. 1999.

Vol. m 61: V. Perlick, Ray Optics, Fermat's Principle, and Applications to General Relativity. X, 220 pages. 2000.

Vol. m 62: J. Berger, J. Rubinstein, Connectivity and Superconductivity. XI, 246 pages. 2000.

Vol. m 63: R. J. Szabo, Ray Optics, Equivariant Cohomology and Localization of Path Integrals. XII, 315 pages. 2000.

Vol. m 64: I. G. Avramidi, Heat Kernel and Quantum Gravity. X, 143 pages. 2000.

Vol. m 65: M. Hénon, Generating Families in the Restricted Three-Body Problem. Quantitative Study of Bifurcations. XII, 301 pages. 2001.

Vol. m 66: F. Calogero, Classical Many-Body Problems Amenable to Exact Treatments. XIX, 749 pages. 2001.

Vol. m 67: A. S. Holevo, Statistical Structure of Quantum Theory. IX, 159 pages. 2001.

Vol. m 68: N. Polonsky, Supersymmetry: Structure and Phenomena. Extensions of the Standard Model. XV, 169 pages. 2001.

Vol. m 69: W. Staude, Laser-Strophometry. High-Resolution Techniques for Velocity Gradient Measurements in Fluid Flows. XV, 178 pages. 2001.

Vol. m 70: P. T. Chruściel, J. Jezierski, J. Kijowski, Hamiltonian Field Theory in the Radiating Regime. VI, 172 pages. 2002.

Vol. m 71: S. Odenbach, Magnetoviscous Effects in Ferrofluids. X, 151 pages. 2002.

Vol. m 72: J. G. Muga, R. Sala Mayato, I. L. Egusquiza (Eds.), Time in Quantum Mechanics. XII, 419 pages. 2002.